Rheology of Complex Fluids

T0143083

Abhijit P. Deshpande • J. Murali Krishnan
P. B. Sunil Kumar
Editors

Rheology of Complex Fluids

 Springer

Editors
Abhijit P. Deshpande
Indian Institute of Technology Madras
Department of Chemical Engineering
Chennai 600036
India
abhijit@iitm.ac.in

P. B. Sunil Kumar
Indian Institute of Technology Madras
Department of Physics
Chennai 600036
India
sunil@physics.iitm.ac.in

J. Murali Krishnan
Indian Institute of Technology Madras
Department of Civil Engineering
Chennai 600036
India
jmk@iitm.ac.in

ISBN 978-1-4899-9727-2   ISBN 978-1-4419-6494-6 (eBook)
DOI 10.1007/978-1-4419-6494-6
Springer New York Dordrecht Heidelberg London

Preface

In complex fluids, the atoms and molecules are organized in a hierarchy of structures from nanoscopic to mesoscopic scales, which in turn make up the bulk material. Consideration of these intermediate scales of organization is essential to understand the behaviour of the complex fluids. These fluids, an important part of what is generally called soft matter, have very complex rheological responses. Many industrial substances encountered in chemical, personal care, food and other processing industries, such as suspensions, colloidal dispersions, emulsions, powders, foams, polymeric liquids and gels, exhibit this complex behaviour. In addition, understanding the behaviour of these complex fluids is also very important in biological systems.

The area of complex fluids and soft matter has been evolving with rapid advances in experimental and computational techniques. The main aim of this book is to introduce these advanced techniques, after a review of fundamental aspects. Since a study of complex fluids involves multidisciplinary tools, contributors with different backgrounds have contributed chapters to this book.

The chapters in this book are based on lectures delivered in the School on Rheology of Complex Fluids, held at Indian Institute of Technology Madras, Chennai during January 4–9, 2010. This school is a part of such series, earlier held at different institutions over the last decade in India. The aim of these schools has been to bring together young researchers and teachers from educational and R&D institutions, and expose them to the basic concepts and research techniques used in the study of rheological behaviour of complex fluids. These schools have been sponsored by the Department of Science and Technology, India.

The book begins with introductory chapters on non-Newtonian fluids, rheological response and fluid mechanics. This is followed by an exposition on how to understand multicomponent and multiphase systems, of which a lot of complex fluids are examples. Analysis of rheological behaviour has been facilitated by experimental and theoretical techniques. The next section of the book gives examples of these in the form of large amplitude oscillatory shear, flow visualizations and stability analysis. The remaining chapters of the book cover application areas of polymers, active fluids and granular materials and their rheology.

IIT Madras, India   Abhijit P. Deshpande
April 2010   J. Murali Krishnan
                                                                     P. B. Sunil Kumar

Acknowledgements

The chapters in this book are based on lectures delivered in the School on Rheology of Complex Fluids, held at Indian Institute of Technology Madras, Chennai during January 4–9, 2010. The school was sponsored by Science and Engineering Research Council, Department of Science and Technology, Government of India. We acknowledge all the participants for active discussions during the school which had helped in shaping the contents of this book.

Devang Khakhar, R P Chhabra, K S Gandhi and Sriram Ramaswamy played an important role in deciding the topics presented here. We thank them for their advice.

We thank R P Chhabra, V Kumaran, K R Rajagopal, S Pushpavanam, Rama Govindarajan, Arti Dua, P Sunthar, Gautam Menon and Mehrdad Massoudi for their contributions. Their help in providing lecture notes, draft chapters and proofreading is acknowledged.

We thank the patient typesetting support provided by VALARDOCS and their team through several modifications of the chapters. Editorial assistance from Santosh was very crucial in bringing the book to its present form. Brett Kurzman and Amanda Davis from Springer helped us in getting through different stages of the book preparation.

Murali Krishnan would like to thank Mary Yvonne Lanzerotti for her encouragement in the earlier stages of this book preparation.

We also acknowledge the support provided by Indian Institute of Technology Madras during different stages of this book preparation.

IIT Madras, India
April 2010

Abhijit P. Deshpande
J. Murali Krishnan
P. B. Sunil Kumar

Contents

Contributors

Basheer Ashraf Ali Research Scholar, Indian Institute of Technology Madras, Chennai 600036, India, ashrafmchem@gmail.com

Rajendra P. Chhabra Department of Chemical Engineering, Indian Institute of Technology Kanpur 208016, India, chhabra@iitk.ac.in

Abhijit P. Deshpande Indian Institute of Technology Madras, Chennai 600036, India, abhijit@iitm.ac.in

Arti Dua Department of Chemistry, Indian Institute of Technology Madras, Chennai 600036, India, arti@iitm.sc.in

Rama Govindarajan Engineering Mechanics Unit, Jawaharlal Nehru Centre for Advanced Scientific Research, Jakkur, Bangalore 560064, India, rama@jncasr.ac.in

V. Kumaran Department of Chemical Engineering, Indian Institute of Science, Bangalore 560 012, India, kumaran@chemeng.iisc.ernet.in

Mehrdad Massoudi U.S. Department of Energy, National Energy Technology Laboratory (NETL), Pittsburgh, PA 15236, USA, massoudi@netl.doe.gov

Gautam I. Menon The Institute of Mathematical Sciences, CIT Campus, Taramani, Chennai 600113, India, menon@imsc.res.in

Subramaniam Pushpavanam Professor, Indian Institute of Technology Madras, Chennai 600036, India, spush@iitm.ac.in

Kumbakonam Ramamani Rajagopal Department of Mechanical Engineering, Texas A&M University, College Station, TX 77845, USA, krajagopal@tamu.edu

Anubhab Roy Engineering Mechanics Unit, Jawaharlal Nehru Centre for Advanced Scientific Research, Jakkur, Bangalore 560064, India, anubhab@jncasr.ac.in

P. Sunthar Department of Chemical Engineering, Indian Institute of Technology Bombay, Mumbai 400076, India, sunthar@che.iitb.ac.in

Acronyms

BSW	Baumgärtel, Schausberger and Winter spectrum
CCD	Charge Coupled Device
CFD	Computational Fluid Dynamics
FENE	Finitely Extensible Nonlinear Elasticity
GNF	Generalized Newtonian Fluids
LASER	Light Amplification by Stimulated Emission of Light Radiation
LAOS	Large Amplitude Oscillatory Shear
LDV	Laser Doppler Velocimetry
PIB	Polyisobutylene
PIV	Particle Image Velocimetry
PM	Parsimonious Model Spectrum
PMFI	Principle of Material Frame-Indifference
PMMA	Polymethylmethacrylate
SAOS	Small Amplitude Oscillatory Shear
SSP	Self-Sustaining Process
VIBGYOR	Violet, Indigo, Blue, Green, Yellow, Orange, Red
WLC	Wormlike Chain

Part I
Background

Chapter 1
Non-Newtonian Fluids: An Introduction

Rajendra P. Chhabra

Abstract The objective of this chapter is to introduce and to illustrate the frequent and wide occurrence of non-Newtonian fluid behaviour in a diverse range of applications, both in nature and in technology. Starting with the definition of a non-Newtonian fluid, different types of non-Newtonian characteristics are briefly described. Representative examples of materials (foams, suspensions, polymer solutions and melts), which, under appropriate circumstances, display shear-thinning, shear-thickening, visco-plastic, time-dependent and viscoelastic behaviour are presented. Each type of non-Newtonian fluid behaviour has been illustrated via experimental data on real materials. This is followed by a short discussion on how to engineer non-Newtonian flow characteristics of a product for its satisfactory end use by manipulating its microstructure by controlling physico-chemical aspects of the system. Finally, we touch upon the ultimate question about the role of non-Newtonian characteristics on the analysis and modelling of the processes of pragmatic engineering significance.

1.1 Introduction

Most low-molecular-weight substances such as organic and inorganic liquids, solutions of low-molecular-weight inorganic salts, molten metals and salts and gases exhibit Newtonian flow characteristics, i.e., at constant temperature and pressure, in simple shear, the shear stress (σ) is proportional to the rate of shear ($\dot{\gamma}$) and the constant of proportionality is the familiar dynamic viscosity (η). Such fluids are classically known as the Newtonian fluids, albeit the notion of flow and of viscosity predates Newton [40]. For most liquids, the viscosity decreases with temperature

R.P. Chhabra (✉)
Department of Chemical Engineering, Indian Institute of Technology, Kanpur 208016, India
e-mail: chhabra@iitk.ac.in

J.M. Krishnan et al. (eds.), *Rheology of Complex Fluids*,
DOI 10.1007/978-1-4419-6494-6_1, © Springer Science+Business Media, LLC 2010

Table 1.1 Values of
viscosity for common fluids
at room temperature

Substance	η(Pa s)
Air	10^{-5}
Water	10^{-3}
Ethyl alcohol	1.2×10^{-3}
Mercury	1.5×10^{-3}
Ethylene glycol	20×10^{-3}
Olive oil	0.1
100% Glycerol	1.5
Honey	10
Corn syrup	100
Bitumen	10^8
Molten glass	10^{12}

and increases with pressure. For gases, it increases with both temperature and pressure [35]. Broadly, higher the viscosity of a substance, more the resistance it presents to flow (and hence more difficult to pump!). Table 1.1 provides typical values of viscosity for scores of common fluids [12]. As we go down in the table, the viscosity increases by several orders of magnitude, and thus one can argue that a solid can be treated as a fluid whose viscosity tends towards infinity, $\eta \to \infty$. Thus, the distinction between a *fluid* and a *solid* is not as sharp as we would like to think! Ever since the formulation of the equations of continuity (mass) and momentum (Cauchy, Navier–Stokes), the fluid dynamics of Newtonian fluids has come a long way during the past 300 or so years, albeit significant challenges especially in the field of turbulence and multiphase flows still remain. During the past 50–60 years, there has been a growing recognition of the fact that many substances of industrial significance, especially of multiphase nature (foams, emulsions, dispersions and suspensions, slurries, for instance) and polymeric melts and solutions (both natural and manmade) do not conform to the Newtonian postulate of the linear relationship between σ and $\dot{\gamma}$ in simple shear, for instance. Accordingly, these fluids are variously known as *non-Newtonian, non-linear, complex* or *rheologically complex fluids*. Table 1.2 gives a representative list of fluids, which exhibit different kinds and varying severity of non-Newtonian flow behaviour [12]. Indeed, so widespread is the non-Newtonian fluid behaviour in nature and in technology that it would be no exaggeration to say that the Newtonian fluid behaviour is an exception rather than the rule! This chapter endeavours to provide a brief introduction to the different kinds of non-Newtonian flow characteristics, their characterization and implications in engineering applications. The material presented herein is mainly drawn from our recent books [11, 12]. The assumptions of material isotropy and incompressibility are implicit throughout our discussion.

Table 1.2 Examples of substances exhibiting non-Newtonian fluid behaviour

Adhesives (wall paper paste, carpet adhesive, for instance)	Food stuffs (Fruit/vegetable purees and concentrates, sauces, salad dressings, mayonnaise, jams and marmalades, ice-cream, soups, cake mixes and cake toppings, egg white, bread mixes, snacks)
Ales (beer, liqueurs, etc.)	Greases and lubricating oils
Animal waste slurries from cattle farms	Mine tailings and mineral suspensions
Biological fluids (blood, synovial fluid, saliva, etc.)	Molten lava and magmas
Bitumen	Paints, polishes and varnishes
Cement paste and slurries	Paper pulp suspensions
Chalk slurries	Peat and lignite slurries
Chocolates	Polymer melts and solutions, reinforced plastics, rubber
Coal slurries	Printing colors and inks
Cosmetics and personal care products (nail polish, lotions and creams, lipsticks, shampoos, shaving foams and creams, toothpaste, etc.)	Pharmaceutical products (creams, foams, suspensions, for instance)
Dairy products and dairy waste streams (cheese, butter, yogurt, fresh cream, whey, for instance)	Sewage sludge
Drilling muds	Wet beach sand
Fire fighting foams	Waxy crude oils

1.2 Classification of Fluid Behaviour

1.2.1 Definition of a Newtonian Fluid

It is useful to begin with the definition of a Newtonian fluid. In simple shear (Fig. 1.1), the response of a Newtonian fluid is characterized by a linear relationship between the applied shear stress and the rate of shear, i.e.,

$$\sigma_{yx} = \frac{F}{A} = \eta \dot{\gamma}_{yx}. \tag{1.1}$$

Figure 1.2 shows experimental results for corn syrup and for cooking oil confirming their Newtonian fluid behaviour; the flow curves pass through the origin and the viscosity values are $\eta = 11.6\,\text{Pa s}$ for corn syrup and $\eta = 64\,\text{mPa s}$ for the cooking oil. Figure 1.1 and (1.1), of course, represent the simplest case wherein there is only one non-zero component of velocity, V_x, which is a function of y. For the general case of three-dimensional flow (Fig. 1.3), clearly there are six shearing and three normal components of the stress tensor, **S**. It is customary to split the total stress into an isotropic part (pressure, p) and a deviatoric part as

$$\mathbf{S} = -p\mathbf{I} + \boldsymbol{\sigma}, \tag{1.2}$$

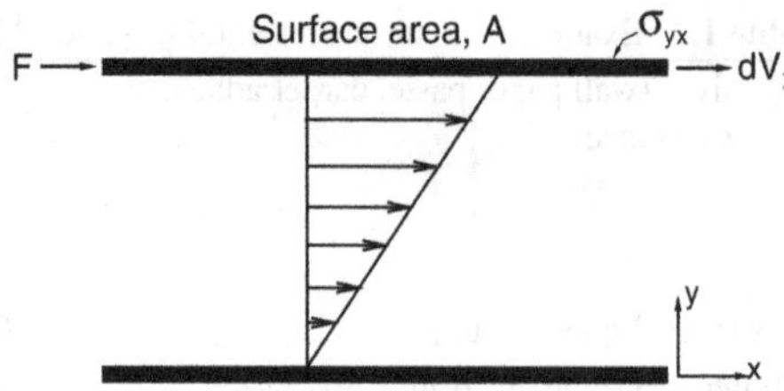

Fig. 1.1 Schematic representation of a unidirectional shearing flow

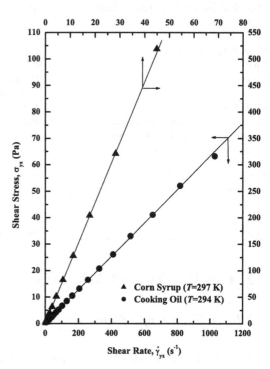

Fig. 1.2 Typical shear stress–shear rate data for two Newtonian fluids

where σ is traceless, i.e., $tr \cdot \sigma = 0$, and pressure is consistent with the continuity equation. The trace-free requirement together with the physical requirement of symmetry $\sigma = \sigma^T$ implies that there are only three independent shear components (off-diagonal elements) and two normal stress differences (diagonal elements) of the deviatoric stress. Thus, in Cartesian coordinates, these are $\sigma_{xy}(= \sigma_{yx})$, $\sigma_{xz}(= \sigma_{zx})$ and $\sigma_{yz}(= \sigma_{zy})$, and the two normal stress differences are defined as

$$\text{Primary normal stress difference, } N_1 = \sigma_{xx} - \sigma_{yy}, \tag{1.3}$$

$$\text{Secondary normal stress difference, } N_2 = \sigma_{yy} - \sigma_{zz}. \tag{1.4}$$

For Newtonian fluids, these components are related linearly to the rate of deformation of tensor components via the scalar viscosity. For instance, the three stress

Fig. 1.3 Stress components
in three-dimensional flow

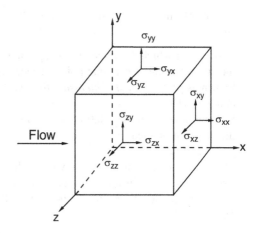

components acting on the x-face (oriented normal to the x-axis) in Fig. 1.3 are
written as follows:

$$\sigma_{xx} = -2\eta \frac{\partial V_x}{\partial x}, \tag{1.5}$$

$$\sigma_{xy} = -\eta \left(\frac{\partial V_x}{\partial y} + \frac{\partial V_y}{\partial x} \right), \tag{1.6}$$

$$\sigma_{xz} = -\eta \left(\frac{\partial V_x}{\partial z} + \frac{\partial V_z}{\partial x} \right). \tag{1.7}$$

Similar sets of equations can be set up for the stress components relevant to the
y- and z-planes. For a Newtonian fluid, in simple shear, $\sigma_{xx} = \sigma_{yy} = \sigma_{zz} = 0$,
because V_x only varies in the y-direction. Thus, the complete definition of a
Newtonian fluid requires it to satisfy the complete Navier–Stokes equations rather
than simply exhibit a constant value of shear viscosity.

1.2.2 Non-Newtonian Fluid Behaviour

The simplest possible deviation from the Newtonian fluid behaviour occurs when
the simple shear data σ–$\dot{\gamma}$ do not pass through the origin and/or does not result in
a linear relationship between σ and $\dot{\gamma}$. Conversely, the apparent viscosity, defined
as $\sigma/\dot{\gamma}$, is not constant and is a function of σ or $\dot{\gamma}$. Indeed, under appropriate
circumstances, the apparent viscosity of certain materials is not only a function of
flow conditions (geometry, rate of shear, etc.), but it also depends on the kinematic
history of the fluid element under consideration. It is convenient, though arbitrary

(and probably unscientific too), to group such materials into the following three categories:

1. Systems for which the value of $\dot{\gamma}$ at a point within the fluid is determined only by the current value of σ at that point, or vice versa, these substances are variously known as *purely viscous, inelastic, time-independent* or *generalized Newtonian fluids (GNF)*;
2. Systems for which the relation between σ and $\dot{\gamma}$ shows further dependence on the duration of shearing and kinematic history; these are called *time-dependent fluids*, and finally,
3. Systems which exhibit a blend of viscous fluid-like behaviour and of elastic solid-like behaviour. For instance, this class of materials shows partial elastic recovery, recoil, creep, etc. Accordingly, these are called *viscoelastic* or *elastico-viscous fluids*.

As noted earlier, the aforementioned classification scheme is quite arbitrary, though convenient, because most real materials often display a combination of two or even all these types of features under appropriate circumstances. For instance, it is not uncommon for a polymer melt to show time-independent (shear-thinning) and viscoelastic behaviour simultaneously and for a china clay suspension to exhibit a combination of time-independent (shear-thinning or shear-thickening) and time-dependent (thixotropic) features at certain concentrations and/or at appropriate shear rates. Generally, it is, however, possible to identify the dominant non-Newtonian aspect and to use it as a basis for the subsequent process calculations. Each type of non-Newtonian fluid behaviour is now dealt with in more detail.

1.3   Time-Independent Fluid Behaviour

As noted above, in simple unidirectional shear, this subset of fluids is characterized by the fact that the current value of the rate of shear at a point in the fluid is determined only by the corresponding current value of the shear stress and vice versa. Conversely, one can say that such fluids have no memory of their past history. Thus, their steady shear behaviour may be described by a relation of the form,

$$\dot{\gamma}_{yx} = f(\sigma_{yx}),\tag{1.8}$$

or, its inverse form,

$$\sigma_{yx} = f^{-1}(\dot{\gamma}_{yx}).\tag{1.9}$$

Depending upon the form of (1.8) or (1.9), three possibilities exist:

1. Shear-thinning or pseudoplastic behavior
2. Visco-plastic behaviour with or without shear-thinning behaviour
3. Shear-thickening or dilatant behaviour.

Fig. 1.4 Qualitative flow curves for different types of non-Newtonian fluids

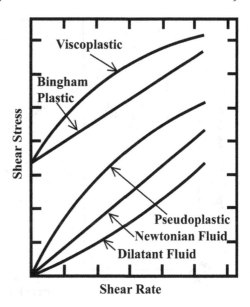

Figure 1.4 shows qualitatively the flow curves (also called rheograms) on linear coordinates for the above-noted three categories of fluid behaviour; the linear relation typical of Newtonian fluids is also included in Fig. 1.4.

1.3.1 Shear-Thinning Fluids

This is perhaps the most widely encountered type of time-independent non-Newtonian fluid behaviour in engineering practice. It is characterized by an apparent viscosity η (defined as $\sigma_{yx}/\dot{\gamma}_{yx}$), which gradually decreases with increasing shear rate. In polymeric systems (melts and solutions), at low shear rates, the apparent viscosity approaches a Newtonian plateau, where the viscosity is independent of shear rate (zero shear viscosity, η_0).

$$\lim_{\dot{\gamma}_{yx} \to 0} \frac{\sigma_{yx}}{\dot{\gamma}_{yx}} = \eta_0. \tag{1.10}$$

Furthermore, polymer solutions also exhibit a similar plateau at very high shear rates (infinite shear viscosity, η_∞), i.e.,

$$\lim_{\dot{\gamma}_{yx} \to \infty} \frac{\sigma_{yx}}{\dot{\gamma}_{yx}} = \eta_\infty. \tag{1.11}$$

In most cases, the value of η_∞ is only slightly higher than the solvent viscosity η_s. Figure 1.5 shows this behaviour in a polymer solution embracing the full spectrum

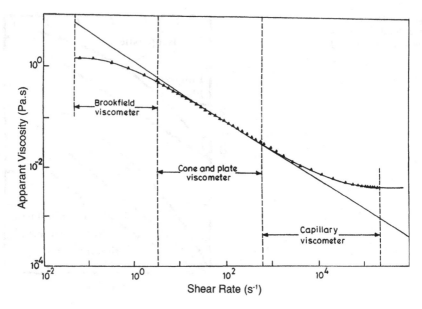

Fig. 1.5 Demonstration of zero-shear and infinite shear viscosities for a polymer solution

of values going from η_0 to η_∞. Obviously, the infinite-shear limit is not seen for polymer melts and blends, or foams or emulsions or suspensions. Thus, the apparent viscosity of a pseudoplastic substance decreases with the increasing shear rate, as shown in Fig. 1.6 for three polymer solutions where not only the values of η_0 are seen to be different in each case, but the rate of decrease of viscosity with shear rate is also seen to vary from one system to another as well as with the shear rate interval considered. Finally, the value of shear rate marking the onset of shear-thinning is influenced by several factors such as the nature and concentration of polymer, the nature of solvent, etc. for polymer solutions and particle size, shape, concentration of solids in suspensions, for instance. Therefore, it is impossible to suggest valid generalizations, but many polymeric systems exhibit the zero-shear viscosity region below $\dot\gamma < 10^{-2}\,\text{s}^{-1}$. Usually, the zero-shear viscosity region expands as the molecular weight of polymer falls, or its molecular weight distribution becomes narrower, as the concentration of polymer in the solution is reduced.

The next question which immediately comes to mind is that how do we approximate this type of fluid behaviour? Over the past 100 years or so, many mathematical equations of varying complexity and forms have been reported in the literature; some of these are straightforward attempts at curve fitting the experimental data $(\sigma - \dot\gamma)$, while others have some theoretical basis (blended with empiricism) in statistical mechanics as an extension of the application of kinetic theory to the liquid state, etc. [9]. While extensive listing of viscosity models is available in several books, for e.g., see Ibarz and Barbosa–Canovas [23] and Govier and Aziz [19], a representative selection of widely used expressions is given here.

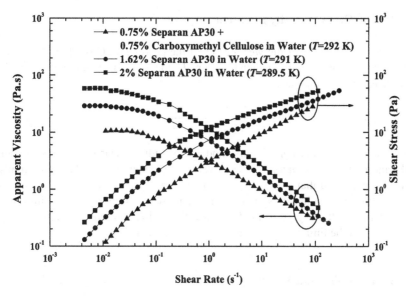

Fig. 1.6 Representative shear stress and apparent viscosity behaviour for three pseudoplastic polymer solutions

1.3.1.1 Power Law or Ostwald de Waele Equation

Often the relationship between shear stress (σ) and shear rate ($\dot{\gamma}$) plotted on log–log coordinates for a shear-thinning fluid can be approximated by a straight line over an interval of shear rate, i.e.,

$$\sigma = m(\dot{\gamma})^n, \tag{1.12}$$

or, in terms of the apparent viscosity,

$$\eta = m(\dot{\gamma})^{n-1}. \tag{1.13}$$

Obviously, $0 < n < 1$ will yield $(d\eta/d\dot{\gamma}) < 0$, i.e., shear-thinning behaviour of fluids is characterized by value of n (power-law index) smaller than unity. Many polymer melts and solutions exhibit the value of n in the range 0.3–0.7 depending upon the concentration and molecular weight of the polymer, etc. Even smaller values of power-law index ($n \sim 0.1-0.15$) are encountered with fine particle suspensions like kaolin-in-water, bentonite-in-water, etc. Naturally, smaller the value of n, more shear-thinning is the material. The other constant, m (consistency index), is a measure of the consistency of the substance.

Although (1.12) or (1.13) offers the simplest approximation of shear-thinning behaviour, it predicts neither the upper nor the lower Newtonian plateau in the limits of $\dot{\gamma} \rightarrow 0$ or $\dot{\gamma} \rightarrow \infty$. Besides, the values of m and n are reasonably constant only over a narrow interval of shear rate range whence one needs to know a priori the likely range of shear rate to be encountered in an envisaged application.

1.3.1.2 The Cross Viscosity Equation

In order to rectify some of the weaknesses of the power-law, Cross [14] presented the following empirical form, which has gained wide acceptance in the literature. In simple shear, it is written as

$$\frac{\eta - \eta_\infty}{\eta_0 - \eta_\infty} = \frac{1}{1 + m(\dot{\gamma})^n}. \tag{1.14}$$

It is readily seen that for $n < 1$, this model also predicts shear-thinning behaviour. Furthermore, the Newtonian limit is recovered here when $m \to 0$. Though initially Cross [14] proposed that $n = 2/3$ was satisfactory for numerous substances, it is now thought that treating it as an adjustable parameter offers significant improvement in terms of the degree of fit [5]. Evidently, (1.14) correctly predicts $\eta = \eta_0$ and $\eta = \eta_\infty$ in the limits of $\dot{\gamma} \to 0$ and $\dot{\gamma} \to \infty$, respectively.

1.3.1.3 The Ellis Fluid Model

While the two viscosity models presented thus far are examples of the form of (1.9), the Ellis model is an illustration of the inverse form, (1.8). In unidirectional simple shear, it is written as

$$\eta = \frac{\eta_0}{1 + \left(\frac{\sigma}{\sigma_{1/2}}\right)^{\alpha-1}}. \tag{1.15}$$

In (1.15), η_0 is the zero-shear viscosity and the remaining two parameters $\sigma_{1/2}$ and $\alpha > 1$ are adjusted to obtain the best fit to a given set of data. Clearly, $\alpha > 1$ yields the decreasing values of shear viscosity with increasing shear rate. It is readily seen that the Newtonian limit is recovered by setting $\sigma_{1/2} \to \infty$. Furthermore, when $(\sigma/\sigma_{1/2}) \gg 1$, (1.15) reduces to the power-law model, (1.12) or (1.13).

1.3.2 Visco-Plastic Fluid Behaviour

This type of non-Newtonian fluid behaviour is characterized by the existence of a threshold stress (called yield stress or apparent yield stress, σ_0), which must be exceeded for the fluid to deform (shear) or flow. Conversely, such a substance will behave like an elastic solid (or flow *en masse* like a rigid body) when the externally applied stress is less than the yield stress, σ_0. Of course, once the magnitude of the external yield stress exceeds the value of σ_0, the fluid may exhibit Newtonian behaviour (constant value of η) or shear-thinning characteristics, i.e., $\eta(\dot{\gamma})$. It therefore stands to reason that, in the absence of surface tension effects, such a material will not level out under gravity to form an absolutely flat free surface. Quantitatively, this type of behaviour can be hypothesized as follows: such a substance at rest consists of three-dimensional structures of sufficient rigidity to resist any external stress less than $|\sigma_0|$ and therefore offers an enormous resistance to flow, albeit it still might deform elastically. For stress levels above $|\sigma_0|$, however, the structure breaks down and the substance behaves like a viscous material. In some cases, the build-up and breakdown of structure has been found to be reversible, i.e., the substance may regain its (initial or somewhat lower) value of the yield stress following a long period of rest.

A fluid with a linear flow curve for $|\sigma| > |\sigma_0|$ is called a Bingham plastic fluid, and is characterized by a constant value of viscosity η_B. Thus, in one-dimensional shear, the Bingham model is written as:

$$\sigma_{yx} = \sigma_0^B + \eta_B \dot{\gamma}_{yx} \qquad\qquad |\sigma_{yx}| > \left|\sigma_0^B\right| \qquad\qquad (1.16a)$$

$$\dot{\gamma}_{yx} = 0 \qquad\qquad |\sigma_{yx}| < \left|\sigma_0^B\right| \qquad\qquad (1.16b)$$

On the other hand, a visco-plastic material showing shear-thinning behaviour at stress levels exceeding $|\sigma_0|$ is known as a yield-pseudoplastic fluid, and its behaviour is frequently approximated by the so-called Herschel–Bulkley fluid model written for 1-D shear flow as follows:

$$\sigma_{yx} = \sigma_0^H + m(\dot{\gamma}_{yx})^n \qquad\qquad |\sigma_{yx}| > \left|\sigma_0^H\right| \qquad\qquad (1.17a)$$

$$\dot{\gamma}_{yx} = 0 \qquad\qquad |\sigma_{yx}| < \left|\sigma_0^H\right| \qquad\qquad (1.17b)$$

Another commonly used viscosity model for visco-plastic fluids is the so-called Casson model, which has its origins in modelling the flow of blood, but it has been found a good approximation for many other substances also [5, 6]. It is written as:

$$\sqrt{|\sigma_{yx}|} = \sqrt{|\sigma_0^C|} + \sqrt{\eta_C\,|\dot{\gamma}_{yx}|} \qquad\qquad |\sigma_{yx}| > \left|\sigma_0^C\right| \qquad\qquad (1.18a)$$

$$\dot{\gamma}_{yx} = 0 \qquad\qquad |\sigma_{yx}| < \left|\sigma_0^C\right| \qquad\qquad (1.18b)$$

While qualitative flow curves for a Bingham fluid and for a yield-pseudoplastic fluid are included in Fig. 1.4, experimental data for a synthetic polymer solution and a meat extract are shown in Fig. 1.7. The meat extract ($\sigma_0 = 17\,\mathrm{Pa}$) conforms to (1.16), whereas the carbopol solution ($\sigma_0 = 68\,\mathrm{Pa}$) shows yield-pseudoplastic behaviour.

Typical examples of yield-stress fluids include blood, yoghurt, tomato puree, molten chocolate, tomato sauce, cosmetics, nail polishes, foams, suspensions, etc. Thorough reviews on the rheology and fluid mechanics of visco-plastic fluids are available in the literature [3, 6].

Finally, before leaving this subsection, it is appropriate to mention here that it has long been a matter of debate and discussion in the literature whether a true yield stress exists or not, e.g., see the trail-blazing paper of Barnes and Walters [4] and the review of Barnes [3] for different viewpoints on this matter. Many workers view the

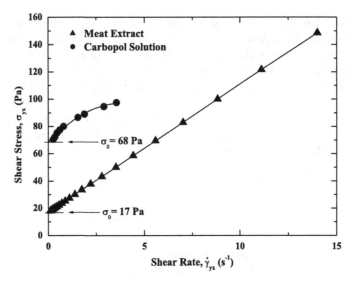

Fig. 1.7 Shear stress–shear rate data for a meat extract and for a carbopol solution displaying Bingham plastic and visco-plastic behaviours, respectively

yield stress in terms of a transition from solid-like behaviour to fluid-like behaviour which manifests itself in terms of an abrupt decrease in viscosity (by several orders of magnitude in many substances) over an extremely narrow range of shear rate [43]. Evidently, the answer to the question whether a substance has a yield stress or not seems to be closely related to the choice of a time scale of observation. In spite of this fundamental difficulty, the notion of an apparent yield stress is of considerable value in the context of engineering applications, especially for product development and design in food, pharmaceutical and health care sectors [3, 36].

1.3.3 Shear-Thickening or Dilatant Behaviour

This class of fluids is similar to pseudoplastic systems in that it shows no yield stress, but its apparent viscosity increases with the increasing shear rate and hence the name *shear-thickening*. Originally, this type of behaviour was observed in concentrated suspensions, and one can qualitatively explain it as follows: at rest, the voidage of the suspension is minimum and the liquid present in the sample is sufficient to fill the voids completely. At low shearing levels, the liquid lubricates the motion of each particle past another thereby minimizing solid–solid friction. Consequently, the resulting stresses are small. At high shear rates, however, the mixture expands (dilates) slightly (similar to that seen in sand dunes) so that the available liquid is no longer sufficient to fill the increased void space and to prevent direct solid–solid contacts (and friction). This leads to the development of much larger shear stresses

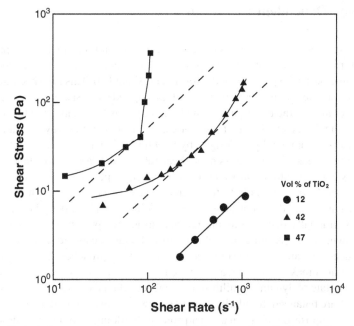

Fig. 1.8 Typical shear stress–shear rate data for TiO_2 suspensions displaying shear-thickening behaviour

than that seen in a pre-dilated sample at low shear rates. This mechanism causes the apparent viscosity $\eta(= \sigma/\dot{\gamma})$ to rise rapidly with the increasing rate of shear. Figure 1.8 shows the representative data for aqueous TiO_2 suspensions of various concentrations [30]. For reference, the lines of slope unity (Newtonian behaviour) are included in this figure. Evidently, these suspensions exhibit both shear-thinning and shear-thickening behaviours over different ranges of shear rate and/or at different concentrations.

Of the time-independent fluids, this sub-class has generated very little interest and hence very few reliable data are available. Indeed, until up to about early 1980s, this type of flow behaviour was considered to be rare, but, however, with the recent growing interest in the handling and processing of systems with high solids loadings, it is no longer so, e.g., see the works of Barnes [1], Boersma et al. [8], Goddard and Bashir [17]. Typical examples of fluids showing shear-thickening behaviour include thick suspensions and pastes of kaolin, TiO_2 and corn flour in water.

The currently available limited information (mostly restricted to simple shear) suggests that it is possible to approximate σ–$\dot{\gamma}$ data for these systems also by the power law model, (1.12), with the power-law index (n) taking on values greater than unity. Notwithstanding the paucity of rheological data on such systems, it is not yet possible to say with confidence whether these materials also display limiting viscosities in the limits of $\dot{\gamma} \rightarrow 0$ and $\dot{\gamma} \rightarrow \infty$.

1.4 Time-Dependent Behaviour

Many substances, notably in food, pharmaceutical and personal care product manufacturing sectors display flow characteristics, which cannot be described by a simple mathematical expression of the form of (1.8) or (1.9). This is so because their apparent viscosities are not only functions of the applied shear stress (σ) or the shear rate ($\dot{\gamma}$), but also of the duration for which the fluid has been subjected to shearing as well as their previous kinematic histories. For instance, the way the sample is loaded into a viscometer, by pouring or by injecting using a syringe, etc. influences the resulting values of shear stress σ or shear rate $\dot{\gamma}$. Similarly, for instance, when materials such as bentonite-in-water, coal-in-water suspensions, red mud suspensions (a waste from alumina industry), cement paste, waxy crude oil, hand lotions and creams are sheared at a constant value of $\dot{\gamma}$ following a long period of rest, their viscosities gradually decrease as the internal structures present are progressively broken down. As the number of such structural linkages capable of being broken down reduces, the rate of change of viscosity with time approaches zero. Conversely, as the structure breaks down, the rate at which linkages can rebuild increases, so that eventually a state of dynamic equilibrium is reached when the rates of buildup and of break down are balanced. Similarly, there are a few systems reported in the literature in which the imposition of external shear promotes building up of internal structures and consequently their apparent viscosities increase with the duration of shearing.

Depending upon the response of a material to shear over a period of time, it is customary to sub-divide time-dependent fluid behaviour into two types, namely, thixotropy and rheopexy (or negative thixotropy). These are discussed in some detail in the next section.

1.4.1 Thixotropic Behaviour

A material is classified as being thixotropic if, when it is sheared at a constant rate, its apparent viscosity $\eta = \sigma/\dot{\gamma}$ (or the value of σ because $\dot{\gamma}$ is constant) decreases with the duration of shearing, as shown in Fig. 1.9 for a red mud suspension containing 59% (by wt) solids [33]. As the value of $\dot{\gamma}$ is gradually increased, the time needed to reach the equilibrium value of σ is seen to drop dramatically. For instance, at $\dot{\gamma} = 3.5 \, s^{-1}$, it is of the order of $\sim 1,500 \, s$, which drops to the value of $\sim 500 \, s$ at $\dot{\gamma} = 56 \, s^{-1}$. Conversely, if the flow curve of such a fluid is measured in a single experiment in which the value of $\dot{\gamma}$ is steadily increased at a constant rate from zero to some maximum value and then decreased at the same rate, a hysteresis loop of the form shown schematically in Fig. 1.10 is obtained. Naturally, for a given material, the height, shape and the area enclosed by the loop depend on the experimental conditions such as the rate of increase/decrease of shear rate, the maximum value of shear rate, and the past kinematic history of the sample. It stands to reason that, the larger the enclosed area, more severe is the time-dependent behaviour of the material under discussion. Evidently, the enclosed area would be zero for a purely

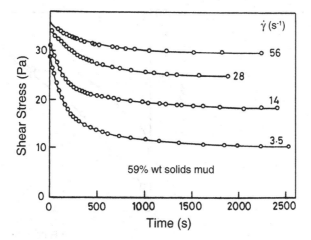

Fig. 1.9 Typical experimental data showing thixotropic behaviour in a red mud suspension [33]

Fig. 1.10 Qualitative shear stress–shear rate behaviour for thixotropic and rheopectic materials

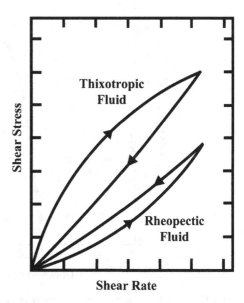

viscous fluid, i.e., no hysteresis effect is expected for time-independent fluids. Data for a cement paste [38] shown in Fig. 1.11 confirm its thixotropic behaviour. Furthermore, in some cases, the breakdown of structure may be reversible, i.e., upon removal of the external shear and following a long period of rest, the fluid may regain (rebuilding of structure) the initial value of viscosity. The data for a lotion shown in Fig. 1.12 illustrate this aspect of thixotropy. Here, the apparent viscosity is seen to drop from $\sim 80\,\mathrm{Pa\,s}$ to $\sim 10\,\mathrm{Pa\,s}$ in about 5–10 s when sheared at $\dot{\gamma} = 100\,\mathrm{s}^{-1}$

Fig. 1.11 Thixotropy
in a cement paste

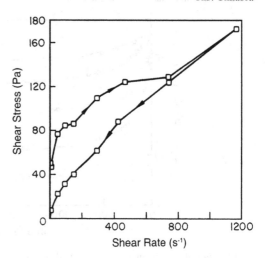

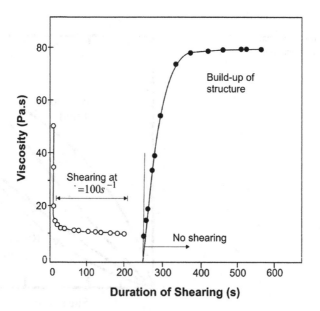

Fig. 1.12 Breakdown and build-up of structure in a proprietary body lotion

and upon removal of the shear, it builds up to almost its initial value in about
50–60 s. Barnes [2] has written a thorough review of thixotropic behaviour encountered in scores of systems of industrial significance.

1.4.2 Rheopectic Behaviour

The relatively few fluids which show the negative thixotropy, i.e., their apparent viscosity (or the corresponding shear stress) increase with time of shearing, are also known as rheopectic fluids. In this case, the hysteresis loop is obviously inverted (Fig. 1.10). As opposed to thixotropic fluids, external shear fosters the build-up of structure in this case. It is not uncommon for the same fluid to display both thixotropy and rheopexy under appropriate combinations of concentration and shear rate. Figure 1.13 shows the gradual onset of rheopexy for a saturated polyester at 60°C [37]. Note that it exhibits time-independent behaviour up to about $\dot{\gamma} \approx 1{,}377\,\text{s}^{-1}$ and the first signature of rheopexy appears only at about $\dot{\gamma} \sim 2{,}755\,\text{s}^{-1}$, which intensifies further with the increasing value of the externally applied shear. Other examples where rheopexy has been observed include suspensions of ammonium oleate and vanadium pentoxide at moderate shear rates, coal–water slurries and protein solutions.

Owing to the frequent occurrence of thixotropic behaviour in a range of industrial settings, much research effort has been devoted to the development of mathematical frameworks to model this type of rheological behaviour [15, 16, 32]. Broadly speaking, three distinct approaches can be discerned, namely, continuum, microstructural and structural kinetics. Within the framework of the continuum approach, existing viscosity models (such as Bingham plastic (1.16), Herschel–Bulkley (1.17), or the Reiner-Rivlin) are amended by postulating the viscosity, yield stress and the other material functions to be functions of time. This, in turn, leads to the power-law consistency and flow behaviour indices to be functions of time. Obviously, the details of micro-structure and changes thereof are completely disregarded in this approach; consequently, it is not at all possible to connect the model parameters

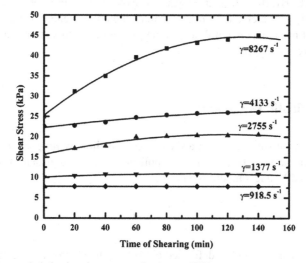

Fig. 1.13 Rheopectic behaviour in a saturated polyester [37]

to the underlying physical processes responsible for the structural changes in the material under shear and for the subsequent build-up of structure upon the removal of shear. On the other hand, the modelling approach based on the consideration of micro-structure requires a detailed knowledge of inter-particle forces, which unfortunately are seldom available for systems of practical significance thereby severely hampering the advancements in this direction. Finally, the thixotropy models based on the structural kinetic arguments hinge on the validity of a single scalar parameter, ξ, which is somehow a measure of the state of the structure in a system. Obviously, it ranges from being zero (completely broken down structure or structure-less!) to being unity (complete buildup of structure). This approach thus comprises two equations: $\sigma - \dot{\gamma}$ relationship for a fixed value of ξ and $\xi - t$ variation, akin to the rate equation for a reversible chemical reaction. This approach is exemplified by the following form of equation due to Houska [22], which has been fairly successful in approximating thixotropic behaviour of scores of systems:

$$\sigma_{yx} = (\sigma_0 + \sigma_{01}) + (m_0 + \xi m_1)\dot{\gamma}^n \qquad (1.19a)$$

$$\dot{\xi} = a(1 - \xi) - b\xi\dot{\gamma}^\varepsilon, \qquad (1.19b)$$

where σ_0 and m_0 are the so-called permanent values of the yield stress and consistency coefficient, respectively; σ_{01} and m_1 are the corresponding time-dependent contributions, which are assumed to be linearly dependent on the current value of ξ. Thus, (1.19a) is valid for a fixed value of the structure parameter ξ. Equation (1.19b) is the kinetic relation which governs the $\xi - t$ relationship. The first term, namely $a(1 - \xi)$, gives the rate of structure buildup (assumed to occur under rest state, i.e., $\dot{\gamma} = 0$) whereas the second term on the right-hand side of (1.19b) gives the rate of breakdown which is a function of both ξ and $\dot{\gamma}$. Altogether, this model contains eight parameters; three (a, b, ε) are kinetic constants and the remaining five are material parameters. Clearly, their evaluation warrants experimental protocols, even in one–dimensional shear, which are far more complex than that needed to characterize the behaviour of time-independent fluids. Some guidelines in this regard are available in the literature [10, 15, 16].

1.5 Viscoelastic Behaviour

For an ideal elastic solid, stress in a sheared state is directly proportional to strain. For tension, the familiar Hooke's law is applicable, and the constant of proportionality is the usual Young's modulus, G, i.e.,

$$\sigma_{yx} = -G\frac{\mathrm{d}x}{\mathrm{d}y} = G(\gamma_{yx}). \qquad (1.20)$$

When an ideal elastic solid is deformed elastically, it regains its original form on removal of the stress. However, if the applied stress exceeds the characteristic yield

Table 1.3 Representative (approximate) values of Young's modulus [29, 36]

Material	Value of G
Glass	70 GPa
Aluminium, copper and alloys	100 GPa
Steel	200 GPa
High modulus oriented fibres	>300 GPa
Concrete	10–20 GPa
Stones	40–60 GPa
Wood	1–10 GPa
Ice	10 GPa
Engineering plastics	5–20 GPa
Leather	1–100 MPa
Rubber	0.1–5 MPa
Polymer and colloidal solutions	1–100 Pa
Dry spaghetti	3 GPa
Carrots	20–40 MPa
Pears	10–30 MPa
Potatoes	6–14 MPa
Peach	2–20 MPa
Raw apples	6–14 MPa
Gelatin gel	0.2 MPa
Banana	0.8–3 MPa

stress of the material, complete recovery will not occur and *creep* will take place – i.e., the solid will have flowed! Table 1.3 presents typical values of the Young's modulus G for a range of materials including metals, plastics, polymer and colloidal solutions and foodstuffs, and these values provide a basis for labeling some of the substances as *soft solids*.

At the other extreme is the Newtonian fluid for which the shearing stress is proportional to the rate of shear, (1.1). Many materials of engineering importance show both elastic and viscous effects under appropriate circumstances. In the absence of thixotropy and rheopexy effects, the material is said to be viscoelastic. Obviously, perfectly viscous flow and perfectly elastic deformation denote the two limiting cases of viscoelastic behaviour. In some materials, only these limiting conditions are observed in practice. Thus, for example, the viscosity of ice and the elasticity of water may generally go unnoticed!! Furthermore, the response of a material is governed not only by its structure, but also by the kinematic conditions it experiences. Therefore, the distinction between a *solid* and a *fluid*, and between an *elastic* and a *viscous* response is to some extent arbitrary and subjective whence is far from being clearcut. Conversely, it is not uncommon for the same material to exhibit viscous fluid-like behaviour in one situation and elastic solid-like behaviour in another situation.

Many materials of pragmatic significance (particularly polymeric melts and solutions, soap solutions, gels, synovial fluid, emulsions, foams, etc.) exhibit viscoelastic behaviour. Thus, for instance, such materials have some ability to store and recover shear energy. One consequence of this type of fluid behaviour is that shearing motion gives rise to stresses (the so-called normal stresses) in the direction normal to that of shear. The resulting normal stresses or normal stress differences N_1 and N_2, defined by (1.3) and (1.4), are also proportional to shear rate in simple shear. Figures 1.14 and 1.15 show representative data on the first and second normal stress

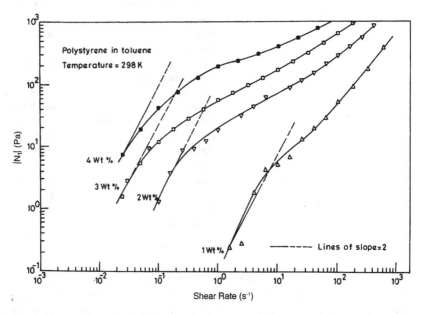

Fig. 1.14 Typical first normal stress difference data for polystyrene-in-toluene solutions [26]

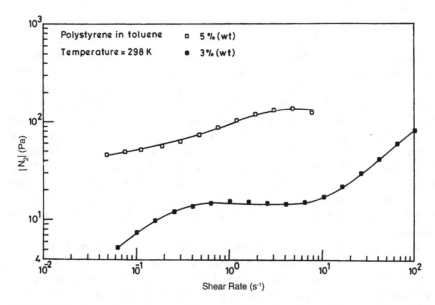

Fig. 1.15 Typical second normal stress difference data for polystyrene-in-toluene solutions [26]

differences for polystyrene-in-toluene solutions at 298 K. Sometimes, it is customary to introduce the first (primary) and second (secondary) normal stress difference
coefficients ψ_1 and ψ_2 defined as follows:

$$\psi_1 = \frac{N_1}{(\dot{\gamma})^2} \tag{1.21a}$$

$$\psi_2 = \frac{N_2}{(\dot{\gamma})^2} \tag{1.21b}$$

Though the actual rates of variation of N_1 and N_2 with shear rate vary from one
system to another, some general observations can be made here. Generally, the rate
of decrease of ψ_1 with $\dot{\gamma}$ is greater than that of the apparent viscosity. At very low
shear rates, N_1 is expected to vary as $\dot{\gamma}^2$, i.e., ψ_1 will approach a constant value in
this limit, as is borne out by some of the data shown in Fig. 1.14. The ratio (N_1/σ)
is often taken to be a measure of the severity of viscoelastic behaviour; specifically,
$(N_1/2\sigma)$ is called the recoverable shear (its values >0.5 are not uncommon for
polymeric systems, which are highly viscoelastic). Generally, experimental determination of N_1 is more difficult than that of shear stress, σ. On the other hand, the
measurements of the second normal stress difference are even more difficult than
that of the first normal stress difference. In most cases, N_2 is typically 10% of N_1
in its magnitude and it is negative. Until the mid 1970s, N_2 was assumed to be zero,
but it is no longer known to be correct.

   Thus, in simple shear, a viscoelastic material is characterized in terms of $N_1 (\dot{\gamma})$,
$N_2 (\dot{\gamma})$ and $\sigma (\dot{\gamma})$; furthermore, the normal stress differences are as such used to
classify a fluid as inelastic $(N_1 \ll \sigma)$ or as viscoelastic $(N_1 \gg \sigma)$.

   So far the discussion has been restricted to the simple unidirectional shearing
flow; now we turn our attention to the two other model flow configurations, namely,
oscillatory shear flow and elongational flow. While the first offers a convenient
method to characterize linear viscoelastic behaviour, the latter denotes idealization
of several industrially important flows.

1.6   Oscillatory Shear Motion

Another common form of flow used to characterize viscoelastic fluids is the so-
called oscillatory shearing motion. It is useful to consider here the response of a
Newtonian fluid and of a Hookean solid to a shear strain which varies sinusoidally
with time as:

$$\gamma = \gamma_m \sin\omega t, \tag{1.22}$$

where γ_m is the amplitude and ω is the frequency of applied strain. For an elastic
Hookean solid, the stress is related linearly to strain, i.e.,

$$\sigma = G\gamma = G\gamma_m \sin\omega t. \tag{1.23}$$

Thus, there is no phase shift between the shear stress and shear strain in this case. On the other hand, for a Newtonian fluid, the shear stress is related to the rate of shear, i.e.,

$$\dot{\gamma} = \frac{d\gamma}{dt} = \gamma_m \omega \cos \omega t = \gamma_m \omega \sin\left(\frac{\pi}{2} + \omega t\right), \tag{1.24}$$

and here

$$\sigma = \eta \dot{\gamma} = \eta \gamma_m \omega \sin\left(\frac{\pi}{2} + \omega t\right) = \sigma_m \sin\left(\frac{\pi}{2} + \omega t\right). \tag{1.25}$$

Obviously in this case, the resulting shear stress is out of phase by $(\pi/2)$ from the applied strain and the stress leads the strain. Thus, the measurement of the phase angle δ, which can vary between zero (purely elastic response) and $(\pi/2)$ (purely viscous response) provides a convenient method of quantifying the level of viscoelasticity of a substance. Needless to add here that small values of δ represent predominantly elastic behaviour, whereas large values of δ correspond to viscous behaviour. For the linear viscoelastic region, one can define the complex viscosity η^* as follows:

$$\eta^* = \eta' + i\eta'', \tag{1.26}$$

where the real and imaginary parts, η' and η'', in turn, are related to the storage (G') and loss (G'') moduli as:

$$\eta'' = \frac{G'}{\omega} \text{ and } \eta' = \frac{G''}{\omega}. \tag{1.27}$$

The storage and loss moduli G' and G'' are defined as:

$$G' = \frac{\sigma_m}{\gamma_m} \cos\delta, \tag{1.28}$$

$$G'' = \frac{\sigma_m}{\gamma_m} \sin\delta. \tag{1.29}$$

Many of the commercially available instruments are equipped for performing oscillating shear tests [28].

1.7 Elongational Flow

This model flow is also known as extensional or stretching flow. In this type of flow, a fluid element is stretched in one or more directions, similar to that encountered in fibre spinning and film blowing. Other examples where this type of flow occurs include coalescence of bubbles and enhanced oil recovery using polymer flooding. There are three forms of elongational flows: uniaxial, biaxial and planar, as shown schematically in Fig. 1.16.

Fibre spinning is an example of uniaxial extension (but the rate of stretching varies along the length of the fibre). Tubular film blowing which entails extrusion

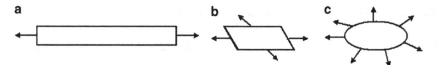

Fig. 1.16   Schematic representation of uniaxial (**a**), biaxial (**b**) and planar (**c**) extensions

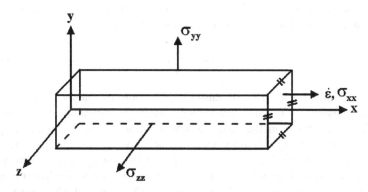

Fig. 1.17   Uniaxial extensional flow

of molten polymers through slit die and pulling the sheet forward and sideways is an illustration of biaxial stretching. Another example is the manufacture of plastic bottles, which are made via extrusion or injection molding, followed by heating and blowing them to the desired size using a high pressure air stream. Due to symmetry, the blowing step is an example of biaxial extension with equal stretching rates in the two directions.

Naturally, the mode of extension influences the way in which the fluid resists deformation and this resistance can be referred to loosely as being quantified in terms of an elongational viscosity, which depends not only upon the rate of stretching but also upon the type of extensional flow. For the sake of simplicity, we consider the uniaxial extension of a fluid element at a constant rate $\dot{\varepsilon}$ in the x-direction as shown in Fig. 1.17. For an incompressible fluid, the volume of the fluid element is conserved, i.e., if it is being stretched in the x-direction at the rate of $\dot{\varepsilon}$, it must shrink in the other directions at the rate of $\dot{\varepsilon}/2$ for an element, which is symmetrical in y- and z- directions. Under these conditions, the velocity vector **V** is given by:

$$\mathbf{V} = \dot{\varepsilon}\, x\hat{i} - (\dot{\varepsilon}/2)\, y\hat{j} - (\dot{\varepsilon}/2)\, z\hat{k}. \tag{1.30}$$

And the rate of elongation $\dot{\varepsilon}$ in the x-direction is given by:

$$\dot{\varepsilon} = \frac{\partial V_x}{\partial x}. \tag{1.31}$$

The extensional viscosity η_E is, in turn, defined as:

$$\eta_E = \frac{\sigma_{xx} - \sigma_{yy}}{\dot{\varepsilon}} = \frac{\sigma_{xx} - \sigma_{zz}}{\dot{\varepsilon}}. \tag{1.32}$$

Early experiments of Trouton [42] on uniaxial elongation by stretching a fibre or a filament of liquid and subsequent studies confirmed that at low elongation rates, the elongational viscosity η_E was three times the corresponding shear viscosity η, and the ratio of the two values is called the Trouton ratio, T_r, i.e.,

$$T_r = \frac{\eta_E}{\eta}. \tag{1.33}$$

The value of 3 for the Trouton ratio for an incompressible Newtonian fluid is valid for all values of $\dot{\varepsilon}$ and $\dot{\gamma}$. By analogy, when this definition of the Trouton ratio, T_r, is extended to include non-Newtonian fluids, one runs into a conceptual difficulty. This is simply due to the fact that for a non-Newtonian fluid, the shear viscosity is a function of the shear rate, $\eta(\dot{\gamma})$, and the elongational viscosity is a function of the rate of stretching, $\eta_E(\dot{\varepsilon})$. Therefore, one needs to adopt a convention for establishing the equivalence between $\dot{\gamma}$ and $\dot{\varepsilon}$. Jones et al. [24] proposed the equivalence as $\dot{\gamma} = \sqrt{3}\dot{\varepsilon}$ and hence the Trouton ratio for a non-Newtonian (incompressible) fluid can now be defined as:

$$T_r = \frac{\eta_E(\dot{\varepsilon})}{\eta(\dot{\varepsilon}\sqrt{3})}. \tag{1.34}$$

Furthermore, Jones et al. [24] proposed that for inelastic isotropic fluids, $T_r = 3$ is applicable for all values of $\dot{\varepsilon}$ and $\dot{\gamma}$, and any departure from the value of 3 can unambiguously be ascribed to the viscoelastic nature of the substance. For an inelastic shear-thinning fluid, this argument predicts tension-thinning in elongation also. On the other hand, values of T_r as large as 1,000 have been documented in the literature for viscoelastic shear-thinning fluids. In other words, such a fluid thins in shear but thickens in tension (strain hardening). Therefore, except in the limits of $\dot{\gamma} \to 0$ and $\dot{\varepsilon} \to 0$, there does not appear to be any simple way enabling the prediction of η_E from a knowledge of η (or vice versa), and the determination of η_E rests entirely on experiments. Figure 1.18 shows representative results on extensional viscosity of a polymer solution for a range of values of $\dot{\varepsilon}$.

While the foregoing discussion shows how a viscoelastic substance displays a blend of fluid-like and solid-like responses under appropriate conditions, the mathematical equations need to be quite complex to adequately describe the behaviour of a real fluid. However, the early attempts are based on the use of mechanical analogues involving different combinations of springs (elastic) and dashpots (viscous) in series, or in parallel, or combinations thereof; three common possibilities are shown in Fig. 1.19. One distinct feature of viscoelastic fluids is the so-called memory effect. For instance, viscous fluids have no memory whereas an ideal elastic solid has a perfect memory as long as the stress is within the linear limit. Thus,

Fig. 1.18 Extensional behaviour of a PIB (Polyisobutylene) solution [41]

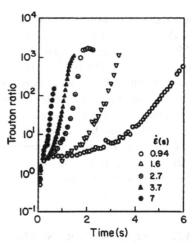

Fig. 1.19 Schematic representation of the Maxwell model (**a**) the Kelvin–Voigt model (**b**) the Burgers' model (**c**)

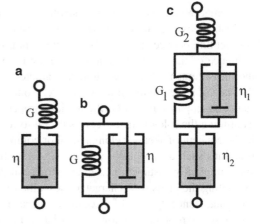

viscoelastic fluids are characterized by using a relaxation time, or a spectrum of re-laxation times, which is roughly a measure of the span of their memory. The relevant dimensionless parameter is the well-known Deborah number, De:

$$De = \frac{\text{Relaxation time of fluid}(\lambda)}{\text{time scale of process}}. \tag{1.35}$$

For the purpose of illustration here, let us consider the flow of a polymer solution (with a relaxation time of 10 ms) in a packed bed of spheres where a fluid element experiences acceleration and deceleration as it flows through the interstices of the bed. For a particle size of 25 mm in an industrial scale packed column and fluid velocity of 250 mm s^{-1}, the time scale of process is of the order of 25/250 $\sim$ 0.1 s, which is much larger than the fluid relaxation time of 10 ms. Therefore, the fluid elements are able to adjust to the changing flow area and one would not expect to see any viscoelastic effects in this case. The corresponding value of the Deborah number is De $= 0.1$. On the other hand, in a laboratory size smaller column comprising $250\,\mu$m diameter spheres and at the same fluid velocity, the time scale of the process is $250 \times 10^{-6}/250 \times 10^{-3} = 10^{-3}$ s, which is much shorter than the fluid relaxation time of 10 ms whence under these conditions, a fluid element is not able to adjust to the changing flow area and hence, viscoelastic effects will manifest [11]. The value of Deborah number in this case is $10 \times 10^{-3}/10^{-3} = 10$. This reinforces the point made earlier that the response of a substance is not governed solely by its structure, but also in conjunction with the type of flow. This section is concluded by noting that De $\rightarrow 0$ represents purely viscous response and De $\rightarrow \infty$ denotes purely elastic response, and most engineering applications occur in between these two limits.

1.8   Origins of Non-Newtonian Behaviour

The foregoing discussion clearly establishes that not only are non-Newtonian characteristics observed in most cases in the so-called structured fluids, but there is also a direct link between the type and extent of non-Newtonian fluid behaviour and the influence of the externally applied stress on the state of the structure. Therefore, the measurement of non-Newtonian characteristics is frequently used to ascertain the state of structure in a fluid. Conversely, one can engineer the structure of a substance to impart the desired rheological properties to a product. However, before examining the role of structure, it is useful to review two key assumptions implicit in the concept of shear or elongational or complex viscosity, and the other material functions such as η, σ_0, N_1, N_2 and G. First, the validity of continuum hypothesis is implicitly assumed, i.e., micro-structural details are deemed unimportant in evaluating the gross flow characteristics, albeit no real fluids are truly structureless continua. Hence, the use of viscosity as a space- and time-averaged physical property poses no problems for low-molecular-weight substances (molecular dimensions $\sim$1–10 nm). Similarly, concentrated polymer solutions or melts, colloidal systems, foams, worm-like micellar systems, etc. possess "micro-structures" of a size approaching 1–2 μm, which can be approximated as a continuum, except during their flow in very fine and/or twisted flow channels. This, in turn, allows the average properties to be defined and assigned values which are not influenced by the dimensions of flow passages. Therefore, as long as the size of micro-structures is much smaller than the characteristic linear scales of the apparatus, one can safely invoke the continuum hypothesis. The second issue concerns the assumption of spatial homogeneity (isotropy?) of a substance so that the space-averaging is meaningful. Finally, as noted here, if all fluids are "structured" to varying extents, what is so special about substances exhibiting non-Newtonian characteristics? The main distinguishing feature is that the structures present in the rheologically complex systems are not only transient in nature, but can also easily be perturbed by the application of relatively low stresses. For instance, the structure of cyclohexane remains unperturbed by the application of stresses up to about 1 MPa. In contrast, the corresponding value is of

Fig. 1.20 Schematics of structures in non-Newtonian dispersions at rest and under shear

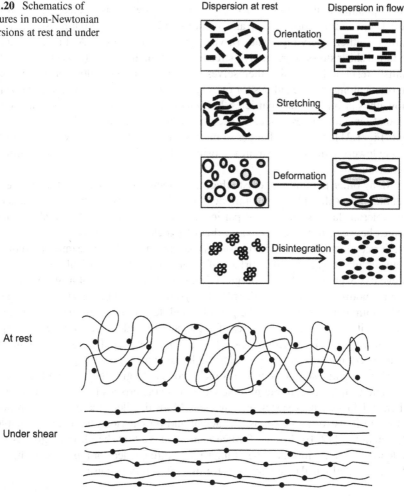

Dispersion at rest   Dispersion in flow

Orientation

Stretching

Deformation

Disintegration

At rest

Under shear

Fig. 1.21 Schematic representation of uncoiling of a macromolecule under the influence of shear

the order of 100 Pa for a polymer of moderate molecular weight and about 200 mPa for a colloidal dispersion containing flow units of the order of 100 nm. It is this degree of ease with which the structure can be perturbed that gives rise to non-Newtonian flow characteristics in a system.

Figures 1.20 and 1.21 show schematically the various types of micro-structures encountered in rheologically complex systems in a rest state (relevant to storage conditions) and how these get perturbed under the action of shear (in a flowing condition). Most systems contain irregularly shaped particles with size distribution (drops and bubbles in emulsions and foams, respectively), or branched and/or entangled molecules in case of polymeric systems, or loosely formed clusters of particles in suspensions, etc. At rest, micro-structured units are oriented randomly corresponding to their minimum energy states. At low levels of shearing, the system resists any deformation or rearrangement by offering a very high resistance by exhibiting either a very high value of viscosity or a yield stress. As the magnitude of external shear stress is gradually increased, the structural units (also known as "flow units") respond by aligning themselves with the direction of flow, or by deforming to orient along the streamlines, or by way of disintegration of aggregates into small flow units or into primary particles. Polymer molecules which are coiled and entangled at low shear rates gradually become disentangled, and finally fully straighten out (Fig. 1.21). All these changes in micro-structures facilitate flow, i.e., these lead to the lowering of their apparent viscosity with shear which leads to shear-thinning behaviour.

Many other possibilities exist which contribute to micro-structural changes depending upon the relative magnitudes of various forces at play. For instance, in sub-micron (large surface area) particle suspensions, the van der Waals attraction forces between particles can cause them to stick to each other. This is responsible for coagulation in colloidal systems (particle size $\sim 1\,\mu m$). Similarly, repulsion between like charges on the surface of a particle produces a repulsive force which prevents coagulation. The rheological behaviour of aqueous kaolin suspensions thus can be modified by adjusting the pH of a system or by adding a surfactant solution. The kaolin consists of plate-like particles and depending upon the type of surface charges, it can form different types of aggregates like edge–face or face–face type in non-flowing conditions (for e.g., see Fig. 1.22). As expected, these two possibilities result in completely different rheological behaviours, e.g., see Fig. 1.23.

The preceding short discussion is included here to give the reader a feel that it is possible to impart desirable non-Newtonian characteristics by tuning the physico-chemical factors in a systematic manner. More detailed treatments of property–structure links for suspensions, surfactant and polymeric systems are available in the literature [12, 16, 20, 21]. Suffice it to say here that the ultimate goal is to be able to predict a priori the type of micro-structure needed for a product to have the desirable rheological characteristics for its satisfactory end use.

1.9 Implications in Engineering Applications

It is natural to ask the question that how does it all impact on the engineering applications involving flow, heat and mass transfer with non-Newtonian fluids? In order to answer this question, for the sake of simplicity, let us restrict our discussion to the flow part only. In principle, one can always set up the equation of continuity and Cauchy's momentum equations (written in their compact form for an incompressible fluid) as follows:

$$\nabla \cdot \mathbf{V} = 0, \tag{1.36}$$

$$\rho \frac{D\mathbf{V}}{Dt} = -\nabla p + \rho g + \nabla \cdot \sigma. \tag{1.37}$$

Fig. 1.22 Possible forms
of agglomerates in kaolin
suspensions

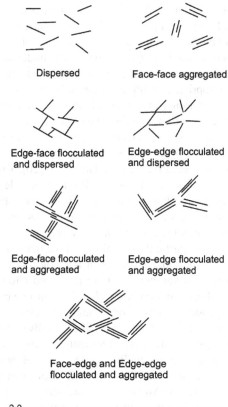

Dispersed   Face-face aggregated

Edge-face flocculated   Edge-edge flocculated
and dispersed   and dispersed

Edge-face flocculated   Edge-edge flocculated
and aggregated   and aggregated

Face-edge and Edge-edge
flocculated and aggregated

Fig. 1.23 Effect of the shape
of agglomerates on the steady
shear behaviour of kaolin
suspensions [18]

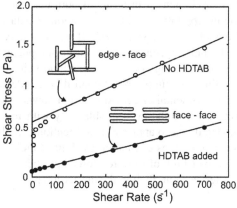

For Newtonian fluids, the deviatoric stress tensor σ is related to the rate of deformation tensor by equations similar to that given by (1.5) to (1.7). Significant research effort has been expended in seeking a similar expression for σ for non-Newtonian fluids which should be able not only to predict shear-dependent viscosity, yield stress, visco-elastic effects in shear and extensional flows, rheopexy and thixotropy but should also satisfy the requirements of frame indifference, material objectivity, etc. [7]. Given the diversity of the materials ranging from homogeneous polymer solutions to liquid crystalline polymers, worm-like micellar solution, suspensions, foams, for instance, this is indeed a tall order to expect that a single constitutive equation will perform satisfactorily under all circumstances for all types of materials. Not withstanding the significant advances made in this field, the choice of an appropriate constitutive relation is (and will continue to be) guided by intuition and by experience in so far to identify the dominant characteristic of the material at hand in conjunction with the type of flow (shear dominated, strongly extensional, mixed, etc.). Critical appraisals of the current state-of-the-art and useful guidelines for the selection of an appropriate expression for σ (constitutive equation) are available in the literature, e.g., see Graessley [20], Kroger [25], Morrison [31], Tanner [39] amongst others. Therefore, if one were able to develop an appropriate constitutive equation and/or to choose one from the existing selection, it is possible to set up the governing differential equations together with suitable boundary conditions, albeit there are situations in which the prescription of boundary conditions is also far from obvious, particularly in flows with a free surface, slip, etc.

Furthermore, even when the non-linear inertial terms (corresponding to zero Reynolds number flow) are neglected altogether in the momentum equation, the resulting equations are still highly non-linear due to the constitutive equation (shear-dependent viscosity, other non-linear effects due to visco-elasticity, etc.) Therefore, except for the simple flows like the fully developed one-dimensional flow in circular and planar ducts, one frequently resorts to numerical solutions, which themselves pose enormous challenges in terms of being highly resource intensive and in terms of acute convergence difficulties thereby breaking down for large values of Deborah and Weissenberg numbers [34]. Finally, experimentalists also confront similar challenges in terms of both the material characterization (rheometry) and the interpretation and representation of data using dimensionless groups, e.g., see Coussot [13] and Macosko [28] for rheometry.

In summary, the analysis of transport phenomena problems of engineering significance involving non-Newtonian fluid behaviour is far more challenging than that entailing the simple Newtonian fluids. Indeed, it is such an easy task to produce an experimental effect using a non-Newtonian fluid in the laboratory, which cannot be explained even qualitatively with the help of the simple Newtonian fluid model. It is also appropriate to mention here that it is not always possible to justify the assumptions of incompressibility (think of foams, gas–liquid dispersions) and isotropy (think of fibre-reinforced plastics, liquid crystalline polymers, nano composites, etc.) implicit in the discussion presented in this introductory chapter.

1.10 Concluding Remarks

In this chapter, consideration has been given to the different types of non-Newtonian characteristics displayed by pseudo-homogeneous mixtures including foams, emulsions, suspensions and pastes, macro-molecular systems (polymer melts and solutions, protein solutions), surfactants (soap solutions), reinforced plastics and polymers in their molten state. The discussion here is restricted primarily to the response of such structured fluids in unidirectional steady shearing motion (with limited reference to oscillatory shear and elongational flows), which leads to the manifestation of shear-thinning, shear-thickening, visco-plastic, thixotropic, rheopectic, viscoelastic characteristics. Each of these is described in some detail supported by representative experimental data on real materials. Qualitative explanation for each type of behaviour is advanced to provide some insights into the nature of underlying physical processes. This, in turn, provides some ideas on how to manipulate the micro-structure of a system to realize desirable non-Newtonian features. Conversely, the measurement and monitoring of viscosity, yield stress, etc. is frequently used to control product quality in food and personal care product sectors, for instance. The chapter is concluded by emphasizing the influence of non-Newtonian flow properties in modelling engineering processes.

References

1. Barnes HA (1989) Review of shear-thickening (dilatancy) in suspensions of non-aggregating solid particles dispersed in Newtonian liquids. J Rheol 33:329–366
2. Barnes HA (1997) Thixotropy – a review. J Non-Newt Fluid Mech 70:1–33
3. Barnes HA (1999) The yield stress – a review or $\pi\alpha\nu\tau\alpha$ $\rho\varepsilon\iota$ everything flows? J Non-Newt Fluid Mech 81:133–178
4. Barnes HA, Walters K (1985) The yield stress myth? Rheol Acta 24:323–326
5. Barnes HA, Hutton JF, Walters K (1989) An introduction to rheology. Elsevier, Amsterdam
6. Bird RB, Dai GC, Yarusso BJ (1983) The rheology and flow of viscoplastic materials. Rev Chem Eng 1:1–83
7. Bird RB, Armstrong RC, Hassager O (1987) Dynamics of polymeric liquids. Vols I and II, 2nd edn. Wiley, New York
8. Boersma WH, Laven J, Stein HN (1990) Shear thickening (dilatancy) in concentrated suspensions. AIChEJ 36:321–332
9. Carreau PJ, De Kee D, Chhabra RP (1997) Rheology of polymeric systems. Hanser, Munich
10. Cawkwell MG, Charles ME (1989) Characterization of Canadian arctic thixotropic gelled crude oils utilizing an eight-parameter model. J Pipelines 7:251–264
11. Chhabra RP (2006) Bubbles, drops and particles in non-Newtonian fluids. CRC, FL
12. Chhabra RP, Richardson JF (2008) Non-Newtonian flow and applied rheology. 2nd edn. Butterworth-Heinemann, Oxford
13. Coussot P (2005) Rheometry of pastes, suspensions and granular materials. Wiley, New York
14. Cross MM (1965) Rheology of non-Newtonian fluids: a new flow equation for pseudoplastic systems. J Colloid Sci 20:417–437
15. Dullaert K, Mewis J (2005) Thixotropy: Build-up and breakdown curves during flow. J Rheol 49:1213–1230
16. Dullaert K, Mewis J (2006) A structural kinetic model for thixotropy. J Non-Newt Fluid Mech 139:21–30
17. Goddard JD,Bashir YM (1990) On Reynolds dilatancy. In: recent developments in structured continua, Longman, London
18. Goodwin JW, Hughes RW (2008) Rheology for chemists: an introduction. The Royal Society of Chemistry, Cambridge
19. Govier GW, Aziz K (1977) The flow of complex mixtures in pipes. Van Nostrand, New York

20. Graessley WW (2004) Polymer liquids and networks: structure and properties. Garland science, New York
21. Han CD (2007) Rheology and processing of polymeric materials. Oxford University Press, New York
22. Houska M (1981) PhD thesis, Czech Technical University, Prague
23. Ibarz A, Barbosa-Canovas GV (2003) Unit operations in food engineering. CRC Press, FL
24. Jones DM, Walters K, Williams PR (1987) On the extensional viscosity of mobile polymer solutions. Rheol Acta 26:20–30
25. Kroger M (2004) Simple models for complex non-equilibrium fluids. Phy Rep 390:453–551
26. Kulicke WM, Wallbaum U (1985) Determination of first and second normal stress differences in polymer solutions in steady shear flow and limitations caused by flow irregularities. Chem Eng Sci 40:961–972
27. Larson RG (1998) The structure and rheology of complex fluids. Oxford University Press, New York
28. Macosko CW (1994) Rheology: principles, measurements and applications. Wiley, New York
29. Malkin AY, Isayev AI (2006) Rheology: Concepts, methods and applications. Chem Tec, Toronto
30. Metzner AB, Whitlock M (1958) Flow behaviour of concentrated (dilatant) suspensions. Trans Soc Rheol 2:239–254
31. Morrison FA (2001) Understanding rheology. Oxford University Press, New York
32. Mujumdar A, Beris AN, Metzner AB (2002) Transient phenomena in thixotropic systems. J Non-Newt Fluid Mech 102:157–178
33. Nguyen QD, Uhlherr PHT (1983) Thixotropic behaviour of concentrated red mud suspensions. Proc 3rd Nat Conf Rheol, Melbourne, 63–67
34. Owens RG, Phillips TN (2002) Computational rheology. Imperial College Press, London
35. Reid RC, Prausnitz JM, Sherwood TK (1977) The properties of gases and liquids. 3rd edn. McGraw-Hill, New York
36. Steffe JF (1996) Rheological methods in food process engineering. Freeman, MI
37. Steg I, Katz D (1965) Rheopexy in some polar fluids and in their concentrated solutions in slightly polar solvents. J Appl Polym Sci 9:3177–3193
38. Struble LJ, Ji X (2001) Handbook of analytical techniques in concrete science and technology. William Andrew, New York
39. Tanner RI (2000) Engineering rheology. 2nd edn. Oxford University Press, London
40. Tanner RI, Walters K (1998) Rheology: an historical perspective. Elsevier, Amsterdam
41. Tirtaatmadja V, Sridhar T (1993) A filament stretching device for measurement of extensional viscosity. J Rheol 37:1081–1102
42. Trouton FT (1906) The coefficient of viscous traction and its relation to that of viscosity. Proc Roy Soc A77:426–440
43. Uhlherr PHT, Guo J, Zhang XM et al (2005) The shear-induced solid–liquid transition in yield stress materials with chemically different structures. J Non-Newt Fluid Mech 125:101–119

Chapter 2
Fundamentals of Rheology

V. Kumaran

Abstract Some important fundamental aspects of fluid mechanics and rheology are discussed. First, we examine the basis for the continuum approximation of fluids, and the formulation of equations in the continuum description. Then, we look at stress-strain rate relationships for simple shear flows of both viscous and viscoelastic fluids. The extension of the description to complex flows, which involves the decomposition of the rate of deformation tensor into fundamental components, is discussed. The conservation laws for mass and momentum, and the constitutive relations in the momentum conservation equation are derived. The simple Maxwell model for polymeric fluids, which uses the bead-spring representation of the polymer chain, is discussed.

2.1 Introduction

Rheology [1] deals with the flow of complex fluids. Fluids are different from solids, because fluids continuously deform when there is an applied stress, as shown in Fig. 2.1b, while solids deform and then stop, as shown in Fig. 2.1a. Solids are said to have an 'elastic' response, and can resist an applied stress, while fluids do not have an elastic response, and deform continuously under stress. The distinction between liquids and gases is less fundamental from a macroscopic point of view, even though they are very different at the molecular level. The responses of both liquids and gases to applied stresses are qualitatively similar, even though gases have a much larger deformation rate for a given applied stress than liquids. The objective of rheology is to determine the fluid flow that would be produced due to applied forces [1]. The applied forces could be of different forms. In channels and pipes, for example, the applied force is due to the pressure difference across the ends of the

V. Kumaran (✉)
Department of Chemical Engineering, Indian Institute of Science,
Bangalore 560 012, India
e-mail: kumaran@chemeng.iisc.ernet.in

J.M. Krishnan et al. (eds.), *Rheology of Complex Fluids*,
DOI 10.1007/978-1-4419-6494-6_2, © Springer Science+Business Media, LLC 2010

Fig. 2.1 Deformation in
solids (**a**) and fluid (**b**)

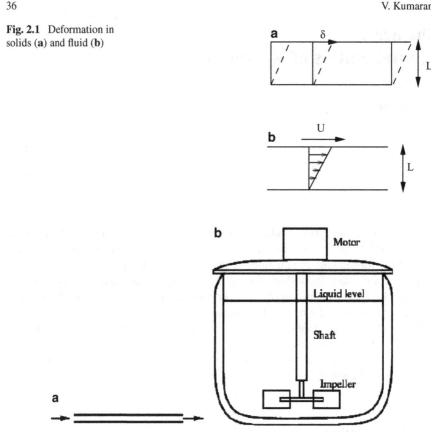

Fig. 2.2 Types of external forces in pipe flow (**a**), reactors (**b**) and flow down inclined plane (**c**)

pipe, as shown in Fig. 2.2a. In reactors, the applied force is due to the rotation of the
impeller, as shown in Fig. 2.2b. The applied force could also be natural in origin,
such as the gravitational force in falling fluid films. An applied force is necessary
for causing fluid flow, because a fluid resists deformation, and this 'fluid friction'
has to be overcome to generate a flow. The objective of rheology is to predict the
flow that would result in a given equipment under the action of applied forces.

The variables used to describe fluid flow [2] are the density, the velocity (which
is a vector and contains three components), the pressure and the stresses. These variables are considered as 'continuum fields', which are continuously varying functions
in space and time. This is not an accurate picture from a microscopic point of view,
of course, because actual fluids consist of molecules which exhibit random motion.
However, this prediction is valid if the length scales of flow, which are macroscopic,
are large compared to the intermolecular separation, and are defined as averages
over volumes, which are small compared to the macroscopic volume under consideration, but large compared to molecular scales. For example, to determine the
density field, consider a volume of fluid ΔV shown in Fig. 2.3a. The total number

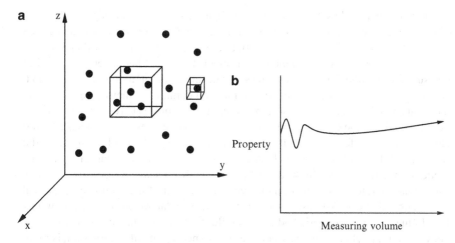

Fig. 2.3 Measuring volume in a molecular fluid (**a**), and dependence of property on measuring volume (**b**)

of particles in the volume is N, and the mass of a molecule is M. The 'density' of the fluid within this volume is defined as

$$\rho = \lim_{\Delta V \to 0} \frac{NM}{\Delta V} \qquad (2.1)$$

If the volume ΔV is of molecular scale, then there will be large fluctuations in the density, because the density will depend on whether a particle is located within the volume or not. As the volume is made larger, but is still small compared to macroscopic scales, the magnitude of the fluctuations will decrease, because the addition or subtraction of a particle will not make much difference in the total number of particles. The density will approach a constant value in this limit, as shown in Fig. 2.3b and this is the value referred to in the limiting process in the above equations. The fluctuation goes as $1/\sqrt{N}$, where N is the number of particles, so the magnitude of the fluctuation decreases rapidly as the number of particles is increased. When the volume is made still larger, there will be an averaging over macroscopic fluctuations in the density and velocity. This is undesirable because we are trying to capture these real fluctuations. Therefore, it is necessary to ensure that the volume taken for averaging is small compared to macroscopic scales, but still contains a large number of molecules. The average velocity, pressure and stress fields are defined in a similar manner. This process for defining macroscopic variables also has a reflection in the computational fluid dynamics method used for solving fluid flow problems. Here, the domain (flow equipment under consideration) is 'discretized' into little volume elements, and the continuum 'fields' are defined on each of these elements. In this case, it is necessary to ensure that the elements are small compared to the macroscopic scales, so that one does not average over real variations in the system, while they are large compared to molecular size.

Flow prediction has two components. The first is a knowledge of how an individual fluid element deforms in response to applied stresses, and this is called the 'model' for the fluid flow. In practice, the response of an individual element has to be measured using a suitable instrument. Usually, 'rheometers' are used to measure the relationship between stress and deformation of an individual element. However, even rheometers make measurement on definite geometries and not on an individual volume element, and so some inference is required to relate the properties of individual volume elements to the rheometer measurements. The models for fluid elements are of different types. If the internal structure of the fluid is simple, in the sense that it does not contain any long chain molecules, colloidal particles, etc., the model for the fluid is usually well described by Newton's law of viscosity. This law states that the rate of deformation of a volume element of the fluid is proportional to the applied stress. All gases, and most low-molecular-weight fluids, follow this law. In the case of polymeric and colloidal fluids, Newton's law of viscosity is not obeyed, and the relationship between the stress and the rate of deformation is more complicated. These fluids are called non-Newtonian fluids. The constitutive relationship between the stress and the strain rate is based on 'rheological models' for the fluid. These could be phenomenological, on the basis of certain invariance properties that have to be followed by fluids. Rheological models could also be based on statistical mechanics, where we start with a microscopic (molecular) description of the fluid, and use averaging to obtain the average properties. Once we have a relationship between the rate of deformation and the applied stresses, it is necessary to determine how the velocity field in the fluid varies with position and time. This is usually determined using a combination of the fluid model, called the constitutive equation, and the conservation equations for the mass and momentum. The law of conservation of mass states that mass can neither be created nor destroyed, and the law of conservation of momentum states that the rate of change of momentum is equal to the sum of the linear applied forces. These are combined with the constitutive relations to provide a set of equations that describe the variation of the fluid density and velocity fields with position and time.

There are two broad approaches for determining the velocity, density, and other dynamical quantities within the equipment of interest from a knowledge of the variation of the properties in space and time. The first is to use analytical techniques, which involve integration of the equations using a suitable mathematical technique. The second is computational, where the differential equations are 'discretized' and solved on the computer. We shall not go into the details of either of these solution techniques here, but merely sketch how the solution is determined for simple problems. The above solution procedure seems simple enough, and one might expect that once the fluid properties are known, the flow properties can easily be calculated with the aid of a suitable computing technique. However, there are numerous complications as listed below:

1. The constitutive properties of the fluid may not be well known. This is common in polymeric and colloidal systems, where the stress depends in a complicated way on the rate of deformation. In such cases, it is necessary to have a good understanding between the constitutive properties of the fluid and the underlying microstructure.

2. The flow may be turbulent. In a 'laminar' flow, the streamlines are smooth and vary slowly with position, so that the deformation of any fluid element resembles the deformation that would be observed in a test equipment which contained only that element. However, when the flow becomes turbulent, the streamlines assume a chaotic form and the flow is highly irregular. Therefore, the flow of an individual volume element does not resemble the deformation that would be observed in a test sample containing only that element. More sophisticated techniques are required to analyse turbulent flows.
3. The flow may actually be too complicated to solve even on a high-speed computer. This is especially true of complicated equipments such as extruders, mixers, fluidised beds, etc. In this case, it is necessary to have alternative techniques to handle these problems.

An important consideration in fluid flow problems is the 'Reynolds number', which provides the ratio of inertial and viscous effects. Fluid 'inertia' refers to the change in momentum due to acceleration, while 'viscous effects' refers to the change in momentum due to frictional forces exerted due to fluid deformation. The Reynolds number is a dimensionless number $Re = (\rho U L/\eta)$, where ρ and η are the fluid density and viscosity, U is the velocity and L is the characteristic length of the flow equipment. The Reynolds number can also be considered as a ratio between 'convective' and 'diffusive' effects, where the 'convective' effect refers to the transport of momentum due to the motion of fluid elements, while the 'diffusive' effect refers to the transport of momentum due to frictional stresses within the fluid. When the Reynolds number is small, viscous effects are dominant and the dominant mechanism of momentum transport is due to diffusion. When the Reynolds number is large, it is expected that viscous effects can be neglected. In flows of complex fluids that we will be concerned with, the Reynolds number is usually sufficiently small that inertial effects can be neglected. Therefore, these flows are mostly dominated by viscosity, and turbulence is not encountered.

2.2 Concept of Viscosity

Viscosity is a measure of the 'fluid friction' or the 'resistance to flow' of a liquid or a gas. This is best illustrated by considering an experiment where the gap between two parallel plates, of area A and separation L, is filled with a fluid (liquid or gas). One of the plates is moved with a velocity U relative to the second stationary plate. It is of interest to determine the force required to move the upper plate at this constant velocity, as a function of the area, the velocity and the separation of the two plates. A typical situation envisaged might be as shown in Fig. 2.4. The two plates are kept at rest, and the fluid is also initially at rest. The top plate is set into motion at time $t = 0$ with a velocity U. At the initial instant, only the fluid very close to the top plate will be in motion, while the rest of the fluid is at rest. In the long time limit, the velocity is a linear function of distance. The shear viscosity is measured in this

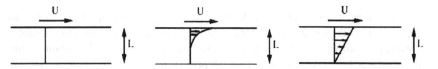

Fig. 2.4 Development of a steady flow in a channel between two plates separated by a distance L when one of the plates is moved with a velocity U

steady state, when the fluid velocity has become a linear function of the distance across the gap. The following can be anticipated quite easily

1. The force required is proportional to the area of the plate A.
2. The force is proportional to the velocity gradient, (U/L).

Since the force required is proportional to the area, it is more convenient to define the 'stress' σ as the (Force/unit area). This can be written as a constant times the strain rate

$$\sigma = \eta U/L. \tag{2.2}$$

The coefficient of proportionality, η, is called the 'shear viscosity' of the fluid. As η becomes larger, the force required for producing a given rate of deformation is larger, and there is more internal friction in the fluid. The unit of viscosity can be deduced from (2.2). The stress, which is the force per unit area, has units of Pascal or (Newton/m^2). The strain rate has units of s^{-1}, since the velocity has units of (m s) and the separation has units of m. Therefore, the viscosity has units of Pa s. A knowledge of the practical range of application of the strain rate is particularly important, since the viscosity depends on the range of the strain rate and other parameters, such as the temperature and the microstructure. In this respect, the viscosity of liquids and gases have very different dependencies on the temperature and strain rate. This difference arises primarily due to the difference in the microstructure.

1. In *gases*, the molecules are separated by distances that are large compared to the molecular diameter, and they interact primarily due to intermolecular collisions. If the velocity scale is small compared to the fluctuating velocity of the molecules, and the length scale is large compared to the distance between the molecules, the viscosity of a gas is independent of the strain rate. It has a strong dependence on the temperature, however, and the viscosity of gases increases proportional to $\sqrt{T}$, where T *is the absolute temperature in Kelvin*.
2. In *liquids*, the separation of molecules is comparable to the molecular length. Due to this, the viscosity has a strong dependence on the temperature, as well as the strain rate. The flow of a liquid takes place due to the sliding of layers of the liquid past each other, and this sliding is easier when the temperature is higher and the molecules have a higher fluctuating velocity. Thus, the viscosity typically decreases with increase in temperature. However, the dependence on the strain rate can be quite complex.

Fig. 2.5 Curve of
$\log\,(d\eta/dT)$ vs. $\log\,(\eta)$ for a
liquid where transport is due
to hopping mechanism

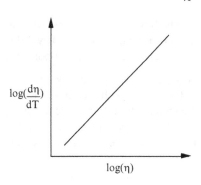

2.2.1 Temperature Dependence

The viscosity of Newtonian liquids decreases with an increase in temperature, according to an Arrhenius relationship. The reason for this kind of relationship is not quite clear. However, it is thought that relative motion between the different molecules in a fluid is due to a type of activated 'hopping' mechanism, and so the viscosity of a liquid also obeys an Arrhenius type behaviour.

$$\eta = A \exp\,(-B/T), \tag{2.3}$$

where T is the absolute temperature (Kelvin), A and B are constants of the liquids. The constant B can be determined by plotting $\log\,(d\eta/dT)$ vs. $\log\eta$, as shown in Fig. 2.5. This curve has a slope of (B/T^2), where T is the absolute temperature. The temperature dependence is stronger for liquids with greater viscosity, and under normal conditions, the thumb rule is that the viscosity decreases by approximately 3% for every degree rise in the temperature. Consequently, it is important to be able to maintain a constant temperature in viscometric measurements. This is of particular importance because the shear of a liquid results in viscous heating, which could change the temperature. Therefore, in a practical measurement, care has to be taken to ensure that sufficient heat is extracted to maintain the system at a steady temperature. The viscosity of liquids increases exponentially with pressure. However, this dependence is not very strong for pressure variations of the order of a few bars, and so is usually ignored in practical applications.

2.2.2 Dependence on Shear Rate

The shear rate dependence of fluids is an important consideration, since many fluids have complex shear rate dependence. Newtonian fluids are characterised by the following behaviour:

1. The viscosity is independent of shear rate.
2. The viscosity is independent of time of shear at a constant shear rate.

3. The 'normal stress differences' are zero.
4. The viscosities measured by different types of deformations, such as uniaxial and biaxial extension, are proportional to each other.

Any liquid showing a deviation from this behaviour is a non-Newtonian liquid. All gases, as well as simple liquids such as water, are Newtonian liquids. Non-Newtonian liquids include all complex liquids such as suspensions, polymeric liquids, etc., which have a complex microstructure where the microstructure depends on the rate of strain of the liquid.

2.2.3 Time Effects

In most fluids which display a shear thinning, there is also a decrease of the viscosity in time at a constant strain rate. This is because of the gradual breakdown of the viscosity under stress, and is similar to the mechanism for shear thickening. There could be a recovery of the structure upon cessation of stress, and the time scale for the recovery gives the typical time for the structures to recover. This type of recovery is typically observed in polymer solutions. There are other cases where the breaking down of bonds is irreversible, such as polymer gels. Here, the viscosity will not recover after the stress has been removed. There are still other cases, such as liquid crystals, where the structures do not break down irreversibly, but the recovery time is long compared to any processing time scale. In this case also there will not be any recovery on the time scale of observation, and the change in structure will appear irreversible. Another widely used concept is the 'yield stress', which is the minimum stress that has to be applied for fluid flow.

However, the concept of yield stress should be used with caution, since the yield stress often depends on the time for which the stress is applied. A material which is rigid for a short duration will flow if stress is continuously applied for a much longer duration. The apparent viscosity of a fluid could also depend on the duration of the applied stress, when the strain rate is maintained a constant. In the case of thixotropic substances, the shear stress decreases with time at a constant strain rate. In contrast, the shear stress of rheopectic substances increases with time at constant shear rate.

2.3 Linear Viscoelasticity

The term 'linear viscoelasticity' refers to the relation between stress and deformation of a material near equilibrium usually when the deformation is small. Deformations are not usually small under conditions of processing or use, and so data obtained by linear viscoelasticity measurements often do not have direct relevance to conditions of use. However, these tests are important for two reasons. First, these tests are easier to carry out and understand than nonlinear measurements,

Fig. 2.6 Displacement due to the stress in a solid

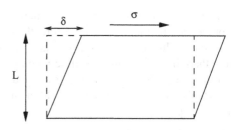

and the parameters obtained from linear viscoelasticity measurements are a reliable reflection of the state of the system. Second, these measurements can also be used in quality control, and any variations in quality can be easily detected by linear viscoelasticity measurements.

The term 'viscoelasticity' implies that the substance under consideration has both viscous and elastic properties. We have already discussed the meaning of the term viscosity. The term elasticity refers to solid materials, and in these cases the stress is a linear function of the *strain γ* (and not the strain rate). If a solid is subjected to a stress, as shown in Fig. 2.6, the solid will undergo deformation up to a certain limit and then stop (provided the stress is less than the breaking point of the solid material, of course). This is in contrast to a liquid, which undergoes continuous deformation when a stress is applied. In Fig. 2.6, if the deformation of the solid is termed δ, then the strain γ is defined as $\gamma = (\delta/L)$, where L is the thickness of the solid. The force required to produce this displacement is proportional to the cross sectional area A, and the stress (force per unit area) is related to the strain field by

$$\sigma = \frac{G\delta}{L}, \tag{2.4}$$

where G is the 'modulus of elasticity'. It is useful to compare the quantity 'strain' used in the definition of elasticity and 'rate of strain' used to define the viscosity. In Fig. 2.6, the velocity of the top plate is the rate of change of the displacement with time,

$$U = \frac{d\delta}{dt}. \tag{2.5}$$

Since the distance between the plates is a constant, the rate of strain is just the time derivative of the strain

$$\dot{\gamma} = \frac{d\gamma}{dt}. \tag{2.6}$$

A 'viscoelastic' fluid is one which has both viscous and elastic properties, i. e. where a part of the stress is due to the strain field and another part is due to the strain rate.

Next, we turn to the term 'linear' in the title 'linear viscoelasticity'. Linear just means that the relationship between the stress and the strain is a linear relationship, so that if the strain is increased by a factor of 2, the stress also increases by a factor of 2. Newton's law of viscosity and (2.4) for the elasticity are linear relationships. However, if we have a term of the form

$$\sigma = K\dot{\gamma}^n, \tag{2.7}$$

it is easy to see that if the strain is increased by a constant factor, the stress does not increase by that factor. Therefore, the relationship is nonlinear. Almost all non-Newtonian fluids have nonlinear relationships between the stress and the strain rate, and therefore they cannot be described by linear viscoelastic theories. However, a linear relationship between the stress and strain could be more complicated than Newton's law of viscosity, and could include higher order time derivatives as well. For example,

$$\left(A_3 \frac{d^3}{dt^3} + A_2 \frac{d^2}{dt^2} + A_1 \frac{d}{dt} + A_0 \right) \sigma = \left(B_3 \frac{d^3}{dt^3} + B_2 \frac{d^2}{dt^2} + B_1 \frac{d}{dt} + B_0 \right) \gamma$$

(2.8)

is a linear relationship between the stress and strain. Relations of the above form do not mean much unless we can relate them to viscous and elastic elements individually. This is the objective of linear viscoelasticity. First, consider viscous and elastic elements separately. They are usually represented diagrammatically as shown in Fig. 2.7.

The stress–strain relationships for these elements are

$$\sigma_E = G\gamma$$
$$\sigma_V = \eta\dot{\gamma}.$$

(2.9)

If these two elements are connected as shown in Fig. 2.8a, the total strain for the viscous and elastic elements is the same, while the total stress is the sum of the stresses due to the viscous and elastic parts.

$$\sigma = G\gamma + \eta\dot{\gamma}.$$

(2.10)

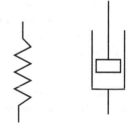

Fig. 2.7 Elastic and viscous elements

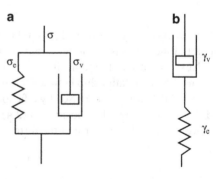

Fig. 2.8 (a) Kelvin model and (b) Maxwell model for the stress–strain relationship

This model is called the 'Kelvin model' for the stress–strain relationship. If a stress is suddenly applied at $t = 0$ and held constant thereafter (step stress), it is easy to show that in the Kelvin model the strain varies as

$$\gamma = (\sigma/G)(1 - \exp(-t/\tau_K)), \tag{2.11}$$

where $\tau_K = (\eta/G)$ is the 'Kelvin time constant'. At long times, the strain attains a constant value (σ/G), which is the strain for a solid with elasticity modulus G. Therefore, the Kelvin model behaves like a solid at long times.

The other way to connect the two elements is as shown in Fig. 2.8b, where the stress transmitted through the two elements is the same, while the total strain is the sum of the strains in the two elements. This type of element is called the 'Maxwell model'. In this case, the stresses in the two elements are

$$\sigma = G\gamma_E$$
$$= \eta\dot{\gamma}_V. \tag{2.12}$$

These two relations can easily be used to obtain the stress–strain relationship

$$\sigma + \tau_M\dot{\sigma} = \eta\dot{\gamma}, \tag{2.13}$$

where $\tau_M = (\eta/G)$ is the Maxwell time constant for this model. If a constant strain rate $\dot{\gamma}$ is applied at time $t = 0$, the evolution of the strain for subsequent times is given by

$$\sigma = \eta\dot{\gamma}(1 - \exp(-t/\tau_M)). \tag{2.14}$$

At long times, we recover the Newton's law for viscosity, and therefore the Maxwell model behaves as a fluid at long times.

It is possible to devise more complicated models with three and four parameters to represent the behaviour of viscoelastic fluids. For example, the Jeffrey's model, shown in Fig. 2.9 is a three-parameter model which in the present notation has a constitutive relation of the form

$$\sigma + \tau_M\dot{\sigma} = \eta(\dot{\gamma} + \tau_j\ddot{\gamma}). \tag{2.15}$$

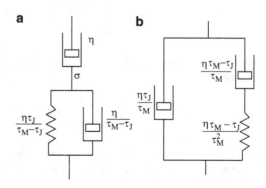

Fig. 2.9 Jeffrey models for stress–strain relationship

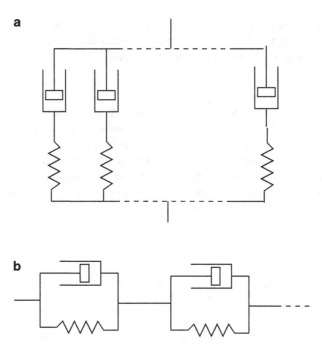

Fig. 2.10 Generalised Maxwell (**a**) and Kelvin (**b**) models for the stress–strain relationships

Other four-parameter models, such as the four-parameter Burgers' model, can also be devised. One can also devise 'generalized Kelvin' and 'generalized Maxwell' models, which have a large number of elements with different time constants (Fig. 2.10).

The stress–strain relationships for these models are complicated, but can be evaluated in principle from the known time constants. For example, if we start with a simple Maxwell model, the stress–strain relationship for a fluid with a varying imposed strain rate $\dot{\gamma}(t)$ is

$$\sigma(t) = \frac{\eta}{\tau_\mathrm{M}} \int_{-\infty}^{t} \exp\left[-(t - t')/\tau_\mathrm{M}\right]\dot{\gamma}(t')\mathrm{d}t'. \tag{2.16}$$

This can easily be extended to the 'generalised Maxwell model' with n Maxwell elements

$$\sigma(t) = \sum_{i=1}^{n} \frac{\eta_i}{\tau_i} \int_{-\infty}^{t} \exp\left[-(t - t')/\tau_i\right]\dot{\gamma}(t')\mathrm{d}t', \tag{2.17}$$

where η_i and τ_i correspond to the ith Maxwell element. A similar relationship can be derived for the 'generalised Kelvin model', where the strain is determined as a function of the applied stress. For a single Kelvin element, the relationship between the stress and strain is

$$\gamma(t) = \frac{1}{\eta} \int_{-\infty}^{t} \exp\left[-(t - t')/\tau_K\right]\sigma(t')dt'. \qquad (2.18)$$

For the generalised Kelvin model, this can easily be extended to provide a relaxation behaviour of the form

$$\gamma(t) = \sum_{i=1}^{n} \frac{1}{\eta_i} \int_{-\infty}^{t} \exp\left[-(t - t')/\tau_i\right]\sigma(t')dt'. \qquad (2.19)$$

The above relationships are quite easy to derive once the models for the viscoelastic fluid are known. However, the basic question is how does one determine the model from a known relationship between the stress and strain. These model relationships are best derived using 'oscillatory' measurements, where a sinusoidal oscillatory strain is imposed on the sample and the stress is recorded. If the strain is sufficiently small (linear response regime), the stress response will also have an oscillatory behaviour with the same frequency (Fig. 2.11).

Most modern rheometers have inbuilt software to carry out the relaxation experiments, and to provide the oscillatory response behaviour as a function of the frequency. The measurement procedure is identical to that for the measurement of viscosity. The only difference in this case is that the imposed angular velocity varies sinusoidally, and the stress is recorded as a function of time. The stress and strain are used to calculate a complex 'shear modulus', and viscometers will usually report the real (storage modulus) and imaginary (loss modulus) parts of the storage modulus. The model parameters can then be determined by the magnitudes of the stress and strain response, and the time lag between the stress and strain.

The stress–strain relationship in this case is usually represented using a complex shear modulus G^*, which is a function of frequency. For example, if the strain imposed is of the form

$$\gamma(t) = \gamma_0 \sin(\omega t). \qquad (2.20)$$

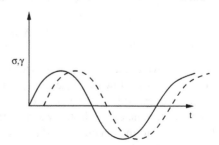

Fig. 2.11 Oscillatory stress and strain

The stress response will be of the form

$$\sigma(t) = G' \gamma_0 \sin(\omega t) + G'' \gamma_0 \cos(\omega t), \tag{2.21}$$

where G' and G'' are the real and imaginary parts of the complex modulus G^*. G' is the 'storage modulus' which gives the response which is in phase with the imposed perturbation, and is related to the elasticity of the material. G'' is the 'loss modulus', which gives the response which is exactly out of phase with the imposed perturbation, and this is related to the viscosity of the material.

The relationship between the complex modulus and the material parameter in the viscoelastic models is best illustrated using the Maxwell model. If the perturbation is of the form given in (2.20), the strain rate is

$$\dot{\gamma} = \frac{d\gamma}{dt} = \gamma_0 \omega \cos(\omega t), \tag{2.22}$$

When this is substituted into the Maxwell model, a first order differential equation is obtained, which can be solved to give

$$\sigma = \frac{\eta \omega \gamma_0}{1 + \omega^2 \tau^2} (\omega \tau \sin(\omega t) + \cos(\omega t)). \tag{2.23}$$

Therefore, the storage and loss moduli are given by

$$\begin{aligned} G' &= \frac{\eta \tau \omega^2}{1 + \omega^2 \tau^2} \\ &= \frac{G \tau^2 \omega^2}{1 + \omega^2 \tau^2} \\ G'' &= \frac{\eta \omega}{1 + \omega^2 \tau^2}. \end{aligned} \tag{2.24}$$

The shapes of the storage and loss moduli 'spectra' are a function of frequency. The storage modulus decreases proportional to ω^2 in the limit of low frequency, and attains a constant value at high frequency. The loss modulus decreases proportional to ω at low frequency, and proportional to ω^{-1} at high frequency. A parameter often quoted in literature is the 'loss tangent' $\tan \delta$, which is defined as

$$\tan \delta = (G''/G'). \tag{2.25}$$

This provides the ratio of viscous to elastic response. This goes to infinity proportional to $(1/\omega)$ at low frequency, indicating that the response is primarily viscous at low frequency. At high frequency, this goes to zero as $(1/\omega)$ indicating that the response is dominated by elasticity at high frequency. The above formalism can easily be extended to the generalised Maxwell model, for example. The storage and loss moduli in this case are related to the time constants and viscosities of the individual Maxwell elements.

$$G' = \sum_{i=1}^{n} \frac{G_i \tau_i^2 \omega^2}{1 + \omega^2 \tau_i^2}$$

$$G'' = \sum_{i=1}^{n} \frac{\eta_i \omega}{1 + \omega^2 \tau_i^2}, \tag{2.26}$$

where G_i, η_i and τ_i are the parameters in the equations for the individual Maxwell elements.

The shape of the storage and loss modulus curves is useful for deducing quantitative information about the relaxation times in the substance. For example, the loss modulus G'' has a peak when $\omega = (1/\tau)$ for a Maxwell fluid with a single time constant. For a generalised Maxwell fluid with two elements, for example, there are two peaks at $\omega = (1/\tau_1)$ and $(1/\tau_2)$ in the loss modulus spectrum. This implies that there are two rheological relaxation mechanisms in the system, and provides their time constants. This could be used to deduce the structure of these fluids. We will examine this a little further while dealing with specific fluids.

2.4 Kinematics

This is the subject of the description of motion, without reference to the forces that cause this motion. Here, we assume that time and space are continuous, and identify a set of particles by specifying their location at a time (x_0, t_0). As the particles move, we follow their positions as a function of time $\{x, t\}$. These positions are determined by solving the equations of motion for the particles, which we have not yet derived. Thus, the positions of the particles at time t can be written as

$$\mathbf{x}(t) = \mathbf{x}(\mathbf{x_0}, t). \tag{2.27}$$

Note that $\mathbf{x}$ and t are independent variables; we can find the location of the particle only if the initial location $\mathbf{x_0}$ is given. Further, we can also express the (initial) position of the particles at time t_0 as a function of their (final) positions at t:

$$\mathbf{x_0}(t_0) = \mathbf{x_0}(\mathbf{x}, t_0). \tag{2.28}$$

2.4.1 Lagrangian and Eulerian Descriptions

The properties of the fluid, such as the velocity and temperature can be expressed in two ways. One is called the 'Lagrangian description', which is a natural extension

Fig. 2.12 Eulerian and
Lagrangian descriptions
of variables in a fluid

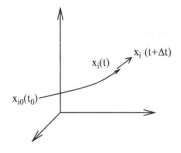

of solid mechanics, where attention is focused on a set of particles in the flow with
initial position $\mathbf{x}_0$, and the evolution of the properties of these particles as they move
through space is determined, as shown in Fig. 2.12. For example, the temperature in
Lagrangian variables is given by

$$T = T(\mathbf{x}_0, t).\tag{2.29}$$

The positions of the particles can be expressed in the Lagrangian variables as noted
below.

$$\mathbf{X} = \mathbf{X}(\mathbf{x}_0, t).\tag{2.30}$$

In the 'Eulerian description', the positions of the properties of the fluid are expressed with reference to positions fixed in space. For example, the velocity and
temperature fields are

$$\mathbf{u} = \mathbf{u}(\mathbf{x}, t)\tag{2.31}$$

$$T = T(\mathbf{x}, t).\tag{2.32}$$

The position vector in Eulerian variables is simply the particle position.

In order to illustrate the Lagrangian viewpoint, let us take a simple example.
Consider a simple linear flow between two flat plates of length L separated by a distance H, with the upper plate moving at a constant velocity V. Neglecting entrance
effects, the fluid velocity is given by

$$u_x = U(z/H)\tag{2.33}$$

$$u_y = 0\tag{2.34}$$

$$u_z = 0.\tag{2.35}$$

In addition, the fluid is being heated on the right and cooled on the left, so that there
is a constant temperature gradient

$$T = T_0 + T_1 x.\tag{2.36}$$

The velocity and temperature as specified above are Eulerian, because they are referenced to a fixed coordinate. To obtain the Lagrangian description, we consider a fluid particle with initial position x_0, y_0, z_0. The velocity of this particle is given by:

$$u_x = U(z_0/H) \tag{2.37}$$

$$u_y = 0 \tag{2.38}$$

$$u_z = 0. \tag{2.39}$$

The velocity of the fluid particle remains constant as the particle moves through the channel, and is independent of time. The particle position is a function of time, however, and is given by

$$x = x_0 + tU(z_0/H) \tag{2.40}$$

$$y = y_0 \tag{2.41}$$

$$z = z_0. \tag{2.42}$$

From the equation for the temperature profile along the length of the channel, we can obtain the Lagrangian form of the temperature profile as well:

$$T = T_0 + T_1 x = T_0 + T_1[x_0 + tU(z_0/H)]. \tag{2.43}$$

This gives the Lagrangian form of the particle position, velocity and temperature for the simple flow that we have considered. For more complex flows, it is very difficult to obtain the Lagrangian description of the particle motion, and this description is not often used.

2.4.2   Substantial Derivatives

Next, we come to the subject of the time derivatives of the properties of a fluid flow. In the Eulerian description, we focus only on the properties as a function of the positions in space. However, note that the positions of the fluid particles are themselves a function of time, and it is often necessary to determine the rate of the change of the properties of a given particle as a function of time, as shown in Fig. 2.12. This derivative is referred to as the Lagrangian derivative or the substantial derivative:

$$\frac{DA}{Dt} = \left\{ \frac{dA}{dt} \right\}\bigg|_{x_0}, \tag{2.44}$$

where A is any general property, and we have explicitly written the subscript to note that the derivative is taken in the Lagrangian viewpoint, following the particle positions in space. For example, consider the substantial derivative of the temperature of a particle as it moves along the flow over a time interval Δt:

$$
\begin{aligned}
\frac{DT}{Dt} &= \lim_{Dt \to 0} \left[\frac{T(\mathbf{x} + \Delta \mathbf{x}, t + \Delta t) - T(\mathbf{x}, t)}{\Delta t} \right] \\
&= \lim_{\Delta t \to 0} \left[\frac{T(\mathbf{x} + \mathbf{u}\Delta t, t + \Delta t) - T(\mathbf{x}, t)}{\Delta t} \right].
\end{aligned}
\tag{2.45}
$$

Taking the limit $\Delta t \to 0$, and using chain rule for differentiation, we obtain

$$
\begin{aligned}
\frac{DT}{dt} &= \frac{\partial T}{\partial t} + u_x \frac{\partial T}{\partial x} + u_y \frac{\partial T}{\partial y} + u_z \frac{\partial T}{\partial z} \\
&= \frac{\partial T}{\partial t} + \mathbf{u}.\nabla T,
\end{aligned}
\tag{2.46}
$$

where

$$
\nabla = \mathbf{e}_x \frac{\partial}{\partial x} + \mathbf{e}_y \frac{\partial}{\partial y} + \mathbf{e}_z \frac{\partial}{\partial z}
\tag{2.47}
$$

is the gradient operator. The relation between the Eulerian and Lagrangian derivatives for the problem just considered can be easily derived. In this example, the Eulerian derivative of the temperature field is zero, because the temperature field has attained steady state. The substantial derivative is given by

$$
\frac{DT}{Dt} = \mathbf{u} \cdot \nabla T = u_x \frac{\partial T}{\partial x} = (Uz/H)T_1.
\tag{2.48}
$$

This can also be obtained by directly taking the time derivative of the temperature field in the Lagrangian description.

2.4.3 Decomposition of the Strain Rate Tensor

The 'strain rate' tensor refers to the relative motion of the fluid particles in the flow. For example, consider a differential volume dV, and two particles located at $\mathbf{x}$ and $\mathbf{x} + \Delta \mathbf{x}$ in the volume, separated by a short distance $\Delta \mathbf{x}$. The velocities of the two particles are $\mathbf{u}$ and $\mathbf{u} + \Delta \mathbf{u}$. The relative velocity of the particles can be expressed in tensor calculus as

$$
d\mathbf{u} = (\nabla \mathbf{u}) \cdot \Delta \mathbf{x},
\tag{2.49}
$$

where $\nabla \mathbf{u}$ is the gradient of the velocity, which is a second-order tensor. In matrix notation, this can be expressed as:

$$
\begin{pmatrix} du_x \\ du_y \\ du_z \end{pmatrix} = \begin{pmatrix} \dfrac{\partial u_x}{\partial x} & \dfrac{\partial u_x}{\partial y} & \dfrac{\partial u_x}{\partial z} \\ \dfrac{\partial u_y}{\partial x} & \dfrac{\partial u_y}{\partial y} & \dfrac{\partial u_y}{\partial z} \\ \dfrac{\partial u_z}{\partial x} & \dfrac{\partial u_z}{\partial y} & \dfrac{\partial u_z}{\partial z} \end{pmatrix} \begin{pmatrix} dx \\ dy \\ dz \end{pmatrix}.
\tag{2.50}
$$

The second-order tensor, $\nabla\mathbf{u}$, can be separated into two components, a symmetric and an antisymmetric component, $\mathbf{S} + \mathbf{A}$, which are given by

$$\mathbf{S} = \frac{1}{2}(\nabla\mathbf{u} + (\nabla\mathbf{u})^{\mathrm{T}}) \tag{2.51}$$

$$\mathbf{A} = \frac{1}{2}(\nabla\mathbf{u} - (\nabla\mathbf{u})^{\mathrm{T}}). \tag{2.52}$$

The antisymmetric part of the strain rate tensor represents rotational flow. Consider a two-dimensional flow field in which the rate of deformation tensor is antisymmetric,

$$\begin{pmatrix} \Delta u_x \\ \Delta u_y \end{pmatrix} = \begin{pmatrix} 0 & -a \\ a & 0 \end{pmatrix} \begin{pmatrix} \Delta x \\ \Delta y \end{pmatrix}. \tag{2.53}$$

The flow relative to the origin due to this rate of deformation tensor is shown in Fig. 2.13a. It is clearly seen that the resulting flow is rotational, and the angular velocity at a displacement Δr from the origin is $a\Delta r$ in the anticlockwise direction.

$$\Delta\mathbf{u} = \frac{1}{2}\Delta\mathbf{x} \times \omega, \tag{2.54}$$

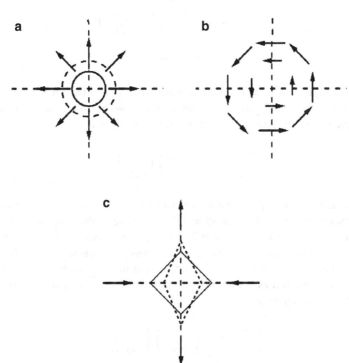

Fig. 2.13 Velocity fields due to the isotropic (**a**), antisymmetric (**b**) and symmetric traceless (**c**) parts of the rate of deformation tensor

where $\times$ is the cross product, and the vorticity ω is

$$\omega = \nabla \times \mathbf{u}. \tag{2.55}$$

The symmetric part of the stress rate tensor can be further separated into two components as follows:

$$\mathbf{S} = \mathbf{E} + \frac{\mathbf{I}}{3}(S_{xx} + S_{yy} + S_{zz}), \tag{2.56}$$

where $\mathbf{E}$ is the symmetric traceless part of the strain tensor, $\mathbf{I}$ is the identity tensor,

$$\mathbf{I} = \begin{pmatrix} 1 & 0 & 0 \\ 0 & 1 & 0 \\ 0 & 0 & 1 \end{pmatrix}, \tag{2.57}$$

and $(1/3)\mathbf{I}(S_{xx} + S_{yy} + S_{zz})$ is the isotropic part. 'Traceless' implies that the trace of the tensor, which is $E_{xx} + E_{yy} + E_{zz}$, is zero. The symmetric traceless part, $\mathbf{E}$, is called the 'extensional strain', while the isotropic part, $(1/3)\mathbf{I}(S_{xx} + S_{yy} + S_{zz})$, corresponds to radial motion. Note that the isotropic part can also be written as

$$\frac{\mathbf{I}}{3}(S_{xx} + S_{yy} + S_{zz}) = \frac{\mathbf{I}}{3}\left(\frac{\partial u_x}{\partial x} + \frac{\partial u_y}{\partial y} + \frac{\partial u_z}{\partial z}\right)$$

$$= \frac{\mathbf{I}}{3}\nabla \cdot \mathbf{u}. \tag{2.58}$$

Thus, the isotropic part of the rate of deformation tensor is related to the divergence of the velocity. The velocity difference between the two neighbouring points due to the isotropic part of the rate of deformation tensor is given by

$$\begin{pmatrix} \Delta u_x \\ \Delta u_y \end{pmatrix} = \begin{pmatrix} s & 0 \\ 0 & s \end{pmatrix}\begin{pmatrix} \Delta x \\ \Delta y \end{pmatrix}. \tag{2.59}$$

The above equation implies that the relative velocity between two points due to the isotropic component is directed along their line of separation, and this represents a radial motion, as shown in Fig. 2.13b. The radial motion is outward if s is positive, and inward if s is negative. The symmetric traceless part represents an 'extensional strain', in which there is no change in density and no solid body rotation. In two dimensions, the simplest example of the velocity field due to a symmetric traceless rate of deformation tensor is

$$\begin{pmatrix} \Delta u_x \\ \Delta u_y \end{pmatrix} = \begin{pmatrix} s & 0 \\ 0 & -s \end{pmatrix}\begin{pmatrix} \Delta x \\ \Delta y \end{pmatrix}. \tag{2.60}$$

The relative velocity of points near the origin due to this rate of deformation tensor is shown in Fig. 2.13c. It is found that the fluid element near the origin deforms in such a way that there is no rotation of the principal axes, and there is no change in the total volume. This type of deformation is called 'pure extensional strain' and is responsible for the internal stresses in the fluid.

The symmetric traceless part of the rate of deformation tensor also contains information about the extensional and compressional axes of the flow. A symmetric tensor of order $n \times n$ has n eigenvalues which are real, and n eigenvectors which are orthogonal. The sum of the eigenvalues of the tensor is equal to the sum of the diagonal elements. Therefore, for the symmetric traceless rate of deformation tensor, the sum of the eigenvalues is equal to zero. This means that, in two dimensions, the two eigenvectors are equal in magnitude and opposite in sign. In three dimensions, the sum of the eigenvalues of the tensor is equal to zero. If $\mathbf{E}_v$ is the column of the normalised eigenvectors of the symmetric traceless tensor $\mathbf{E}$,

$$
\mathbf{E}_v = \left(\left(\mathbf{e}_{v1} \right) \left(\mathbf{e}_{v2} \right) \left(\mathbf{e}_{v3} \right) \right),
\tag{2.61}
$$

and if the eigenvalues are λ_1, λ_2 and λ_3, then the tensor can be written as

$$
\mathbf{E}_v^{-1} \mathbf{E} \mathbf{E}_v = \begin{pmatrix} \lambda_1 & 0 & 0 \\ 0 & \lambda_2 & 0 \\ 0 & 0 & \lambda_3 \end{pmatrix}.
\tag{2.62}
$$

Also, the eigenvectors satisfy the equation $\mathbf{E}_v^{-1} = \mathbf{E}_v^{\mathsf{T}}$. Equation (2.62) can be interpreted as follows. The difference in velocity between two neighbouring locations due to the deviatoric part of the rate of deformation tensor is

$$
\begin{pmatrix} \Delta u_x \\ \Delta u_y \\ \Delta u_z \end{pmatrix} = \mathbf{E} \begin{pmatrix} \Delta x \\ \Delta y \\ \Delta z \end{pmatrix}.
\tag{2.63}
$$

This can be rewritten, using (2.62), as

$$
\begin{pmatrix} \Delta u_x \\ \Delta u_y \\ \Delta u_z \end{pmatrix} = \mathbf{E}_v \begin{pmatrix} \lambda_1 & 0 & 0 \\ 0 & \lambda_2 & 0 \\ 0 & 0 & \lambda_3 \end{pmatrix} \mathbf{E}_v^{-1} \begin{pmatrix} \Delta x \\ \Delta y \\ \Delta z \end{pmatrix}.
\tag{2.64}
$$

If we define a new co-ordinate system

$$
\begin{pmatrix} \Delta x' \\ \Delta y' \\ \Delta z' \end{pmatrix} = \mathbf{E}_v^{-1} \begin{pmatrix} \Delta x \\ \Delta y \\ \Delta z \end{pmatrix},
\tag{2.65}
$$

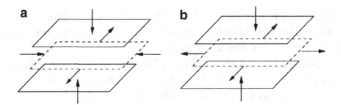

Fig. 2.14 (a) Biaxial extension and (b) uniaxial extension

and velocity

$$\begin{pmatrix} \Delta u'_x \\ \Delta u'_y \\ \Delta u'_z \end{pmatrix} = \mathbf{E}_v^{-1} \begin{pmatrix} \Delta u_x \\ \Delta u_y \\ \Delta u_z \end{pmatrix}, \tag{2.66}$$

then the equation for the velocity difference between two neighbouring locations becomes

$$\begin{pmatrix} \Delta u'_x \\ \Delta u'_y \\ \Delta u'_z \end{pmatrix} = \begin{pmatrix} \lambda_1 & 0 & 0 \\ 0 & \lambda_2 & 0 \\ 0 & 0 & \lambda_3 \end{pmatrix} \begin{pmatrix} \Delta x' \\ \Delta y' \\ \Delta z' \end{pmatrix}. \tag{2.67}$$

The operations in (2.65) and (2.66) are just equivalent to rotating the co-ordinate system. Equation (2.67) indicates that in this rotated co-ordinate system, the differences in velocity are directed along the axes of the rotated co-ordinate system. Since the sum of the three eigenvalues is zero, there are three possibilities for the signs of the three eigenvalues. If two of the eigenvalues are positive and one is negative, then there is extension along two axes and compression along one axis, as shown in Fig. 2.14a. This is called biaxial extension. If one of the eigenvalues is positive and the other two are negative, then there is compression along two axes and extension along one axis, as shown in Fig. 2.14b. This is called uniaxial extension. If one of the eigenvalues is zero, then there is extension along one axis, compression along the second and no deformation along the third axis. This is called plane extension, as shown in Fig. 2.13c.

2.5 Conservation of Mass

The conservation of mass simply states that mass cannot be created or destroyed. Therefore, for any volume of fluid,

$$\begin{pmatrix} \text{Rate of mass} \\ \text{accumulation} \end{pmatrix} = \begin{pmatrix} \text{Rate of mass} \\ \text{IN} \end{pmatrix} - \begin{pmatrix} \text{Rate of mass} \\ \text{OUT} \end{pmatrix}. \tag{2.68}$$

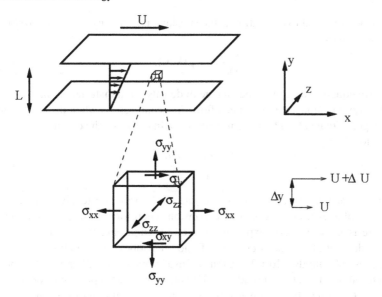

Fig. 2.15 Control volume used for calculating mass and momentum balance

Consider the volume of fluid shown in Fig. 2.15. This volume has a total volume $\Delta x \Delta y \Delta z$, and it has six faces. The rate of mass in through the face at x is $(\rho u_x)|_x \Delta y \Delta z$, while the rate of mass out at $x + \Delta x$ is $(\rho u_x)|_{x+\Delta x} \Delta y \Delta z$. Similar expressions can be written for the rates of mass flow through the other four faces. The total increase in mass for this volume is $(\partial \rho / \partial t) \Delta x \Delta y \Delta z$. Therefore, the mass conservation equation states that

$$\Delta x \Delta y \Delta z \frac{\partial \rho}{\partial t} = \Delta y \Delta z [(\rho u_x)|_x - (\rho u_x)|_{x+\Delta x}]$$
$$+ \Delta x \Delta z [(\rho u_y)|_y - (\rho u_y)|_{y+\Delta y}]$$
$$= \Delta x \Delta y [(\rho u_z)|_z - (\rho u_z)|_{z+\Delta z}]. \quad (2.69)$$

Dividing by $\Delta x \Delta y \Delta z$, and taking the limit as these approach zero, we get

$$\frac{\partial \rho}{\partial t} = -\left(\frac{\partial(\rho u_x)}{\partial x} + \frac{\partial(\rho u_y)}{\partial y} + \frac{\partial(\rho u_z)}{\partial z} \right). \quad (2.70)$$

The above equation can often be written using the substantial derivative

$$\frac{\partial \rho}{\partial t} + \left(u_x \frac{\partial \rho}{\partial x} + u_y \frac{\partial \rho}{\partial y} + u_z \frac{\partial \rho}{\partial z} \right) = -\rho \left(\frac{\partial u_x}{\partial x} + \frac{\partial u_y}{\partial y} + \frac{\partial u_z}{\partial z} \right). \quad (2.71)$$

The left side of the above equation is the substantial derivative, while the right side can be written as

$$\frac{D\rho}{Dt} = -\rho(\nabla.\mathbf{u}). \tag{2.72}$$

The above equation describes the change in density for a material element of fluid, which is moving along with the mean flow. A special case is when the density does not change, so that $(D\rho/Dt)$ is identically zero. In this case, the continuity equation reduces to

$$\frac{\partial u_x}{\partial x} + \frac{\partial u_y}{\partial y} + \frac{\partial u_z}{\partial z} = 0. \tag{2.73}$$

This is just the isotropic part of the rate of deformation tensor, which corresponds to volumetric compression or expansion. Therefore, if this is zero, it implies that there is no volumetric expansion or compression, and if mass is conserved, then the density has to be a constant. Fluids that obey this condition are called 'incompressible' fluids. Most fluids that we use in practical applications are incompressible fluids; in fact, all liquids can be considered incompressible for practical purposes. Compressibility effects only become important in gases when the speed of the gas approaches the speed of sound, $332 \, \mathrm{m \, s^{-1}}$.

2.6 Conservation of Linear Momentum

The conservation of linear momentum is a consequence of Newton's second law of motion, which states that the rate of change of momentum in a differential volume is equal to the total force acting on it. The momentum in a volume $V(t)$ is given by

$$\mathbf{P} = \int_{V(t)} dV \, \rho \mathbf{u}, \tag{2.74}$$

and Newton's second law states that the rate of change of momentum in a moving differential volume is equal to the sum of the applied forces,

$$\frac{D\mathbf{P}_i}{Dt} = \frac{d}{dt} \int_{V(t)} dV \, (\rho \mathbf{u}) = \text{Sum of forces}. \tag{2.75}$$

Here, note that the momentum conservation equation is written for a moving fluid element. In a manner similar to the mass conservation equation, the integral over a moving fluid element can be written in an Eulerian reference frame as

$$\int_{V(t)} dV \left(\frac{\partial \rho \mathbf{u}}{\partial t} + \nabla.(\rho \mathbf{uu}) \right) = \text{Sum of forces}. \tag{2.76}$$

The forces acting on a fluid are usually of two types. The first is the body forces, which act on the bulk of the fluid, such as the force of gravity, and the surface forces, which act at the surface of the control volume. If $\mathbf{F}$ is the body force per unit mass of the fluid, and $\mathbf{R}$ is the force per unit area acting at the surface, then the momentum conservation equation can be written as

$$\int_{V(t)} dV \left(\frac{\partial(\rho\mathbf{u})}{\partial t} + \nabla.(\rho\mathbf{uu}) \right) = \int_{V(t)} dV\mathbf{F} + \int_S dS\mathbf{R}, \qquad (2.77)$$

where S is the surface of the volume $V(t)$.

To complete the derivation of the equations of motion, it is necessary to specify the form of the body and surface forces. The form of the body forces is well known – the force due to gravity per unit mass is just the gravitational acceleration g, while the force due to centrifugal acceleration is given by ωR^2, where ω is the angular velocity and R is the radius. The form of the surface force is not specified a-priori however, and the 'constitutive equations' are required to relate the surface forces to the motion of the fluid. But before specifying the nature of the surface forces, one can derive some of the properties of the forces using symmetry considerations.

The force acting on a surface will, in general, depend on the orientation of the surface, and will be a function of the unit normal to the surface, in addition to the other fluid properties:

$$\mathbf{R} = \mathbf{R}(\mathbf{n}). \qquad (2.78)$$

We can use symmetry considerations to show that the force is a linear function of the unit normal to the surface. If we consider a surface dividing the fluid at a point P, then the forces acting on the two sides of the surface have to be equal according to Newton's third law. Since the unit normals to the two sides are just the negative of each other, this implies that

$$\mathbf{R}(-\mathbf{n}) = -\mathbf{R}(\mathbf{n}). \qquad (2.79)$$

This suggests that the surface force could be a linear function of the normal:

$$\mathbf{R} = \mathbf{T}.\mathbf{n}, \qquad (2.80)$$

where $\mathbf{T}$ is a 'second-order tensor'. That this is the case can be shown using Cauchy's theorem. Consider a tetrahedron with three surfaces located along the three co-ordinate axes, with lengths Δx, Δy and Δz, as shown in Fig. 2.16. In the limit as three lengths go to zero, the density and velocity in the region can be taken as constants. The integral momentum conservation equation then becomes

$$\left(\frac{\partial(\rho\mathbf{u})}{\partial t} + \nabla.(\rho\mathbf{uu}) \right) \Delta V$$
$$= \mathbf{F}\Delta V + \mathbf{R}(\mathbf{n})\Delta S + \mathbf{R}(-\mathbf{e}_x)\Delta S_x + \mathbf{R}(-\mathbf{e}_y)\Delta S_y + \mathbf{R}(-\mathbf{e}_z)\Delta S_z,$$
$$(2.81)$$

where $\mathbf{e}_x$, $\mathbf{e}_y$ and $\mathbf{e}_z$ are the unit vectors in the three directions, S_x, S_y and S_z are the surfaces perpendicular to $\mathbf{e}_x$, $\mathbf{e}_y$ and $\mathbf{e}_z$, respectively, and $\mathbf{R}(\mathbf{n})$ is the force acting

Fig. 2.16 Forces acting on a
tetrahedral control volume

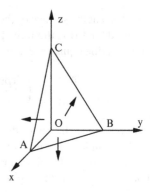

at the surface with outward unit normal **n**. In the limit as the lengths of the sides
$\Delta x, \Delta y, \Delta z$ go to zero, the inertial terms and the body forces, which are proportional to the volume $\Delta x \Delta y \Delta z$, become much smaller than the terms proportional to
the surface area, and therefore,

$$\mathbf{R}(\mathbf{n})\Delta S = \mathbf{R}(\mathbf{e}_x)\Delta S_x + \mathbf{R}(\mathbf{e}_y)\Delta S_y + \mathbf{R}(\mathbf{e}_z)\Delta S_z. \tag{2.82}$$

The areas along the three axes are related to the area of the triangular plane as

$$\Delta S_x = n_x \Delta S \tag{2.83}$$
$$\Delta S_y = n_y \Delta S \tag{2.84}$$
$$\Delta S_z = n_z \Delta S. \tag{2.85}$$

Using this relation, we get

$$\mathbf{R}(\mathbf{n}) = n_x \mathbf{R}(\mathbf{e}_x) + n_y \mathbf{R}(\mathbf{e}_y) + n_z \mathbf{R}(\mathbf{e}_z). \tag{2.86}$$

This is exactly of the same form as

$$\mathbf{R}(\mathbf{n}) = \mathbf{T}.\mathbf{n}. \tag{2.87}$$

In matrix form, this can be written as

$$\begin{pmatrix} R_x \\ R_y \\ R_z \end{pmatrix} = \begin{pmatrix} T_{xx} & T_{xy} & T_{xz} \\ T_{yx} & T_{yy} & T_{yz} \\ T_{zx} & T_{zy} & T_{zz} \end{pmatrix} \begin{pmatrix} n_x \\ n_y \\ n_z \end{pmatrix}. \tag{2.88}$$

The tensor **T** is called the 'stress tensor', and it has nine components. The component T_{xx} is the force per unit area in the x direction acting at a surface whose unit
normal is in the x direction, the component T_{xy} is the force per unit area in the
x direction acting at a surface whose unit normal is in the y direction, and so on.

Therefore, the tensor has two physical directions associated with it; the first is the direction of the force and the second is the direction of the unit normal to the surface at which the stress is measured.

Using the stress tensor, the conservation equation is given by

$$\int_{V(t)} dV \left(\frac{\partial (\rho \mathbf{u})}{\partial t} + \nabla.(\rho \mathbf{u}\mathbf{u}) \right) = \int_{V(t)} dV \, \mathbf{F} + \int_S dS \, \mathbf{T}.\mathbf{n}. \tag{2.89}$$

This can be further simplified using the divergence theorem:

$$\int_{V(t)} dV \left(\frac{\partial (\rho \mathbf{u})}{\partial t} + \nabla.(\rho \mathbf{u}\mathbf{u}) \right) = \int_{V(t)} dV \, \mathbf{F} + \int_S dV \, \nabla.\mathbf{T}. \tag{2.90}$$

Since the above equation is valid for any differential volume in the fluid, the integrand must be equal to zero:

$$\left(\frac{\partial (\rho \mathbf{u})}{\partial t} + \nabla.(\rho \mathbf{u}\mathbf{u}) \right) = \mathbf{F} + \nabla.\mathbf{T}. \tag{2.91}$$

The above equation can be simplified using the mass conservation equation:

$$\rho \left(\frac{\partial \mathbf{u}}{\partial t} + \mathbf{u}\nabla.\mathbf{u} \right) = \mathbf{F} + \nabla.\mathbf{T}. \tag{2.92}$$

An additional property of the stress tensor, which is obtained from the angular momentum equation, is that the stress tensor is symmetric. That is, $T_{xy} = T_{yx}$, $T_{xz} = T_{zx}$ and $T_{yz} = T_{zy}$ in (2.88).

2.7 Constitutive Equations for the Stress Tensor

Just as we had earlier separated the strain rate tensor into an antisymmetric traceless part, a symmetric part and an isotropic part, it is conventional to separate the stress tensor into an isotropic part and a symmetric traceless 'deviatoric' part:

$$\mathbf{T} = -p\mathbf{I} + \boldsymbol{\sigma}, \tag{2.93}$$

where p is the pressure in the isotropic pressure in the fluid, $p = -(1/3)(T_{xx} + T_{yy} + T_{zz})$, and $\boldsymbol{\sigma}$ is the second-order 'deviatoric' stress tensor which is traceless, $(\sigma_{xx} + \sigma_{yy} + \sigma_{zz}) = 0$. In the absence of fluid flow, the deviatoric part of the stress tensor $\boldsymbol{\sigma}$ becomes zero, and the pressure field is related to the local density of the system by a thermodynamic equation of state.

The shear stress, $\boldsymbol{\sigma}$, is a function of the fluid velocity. However, the stress cannot depend on the fluid velocity itself, because the stress has to be invariant under a 'Galilean transformation' i.e., when the velocity of the entire system is changed by a constant value. Therefore, the stress has to depend on the gradient of the fluid velocity. In a 'Newtonian fluid', we make the assumption that the stress is a linear function of the velocity gradient. In general, the linear relation can be written as

$$\sigma = \eta \cdot \nabla \mathbf{u}, \tag{2.94}$$

where η is a fourth order tensor. Here $\Delta \mathbf{u}$ is not a frame invariant measure. This tensor is a property of the fluid.

In an isotropic fluid, there is no preferred direction in space, and therefore, the tensor η should be independent of direction. In this case, it can be shown that the relationship between the stress and strain rate is of the form

$$\sigma = \eta(\nabla \mathbf{u} + (\nabla \mathbf{u})^{\mathrm{T}} - (2/3)\mathbf{I}\nabla.\mathbf{u}), \tag{2.95}$$

where η is the 'coefficient of viscosity' of the fluid. This can also be expressed in terms of the rate of strain tensor:

$$\sigma = 2\eta\mathbf{E}, \tag{2.96}$$

where $\mathbf{E}$ is the symmetric traceless part of the rate of deformation tensor.

The above constitutive equation was derived for a Newtonian fluid with the assumption that the shear stress is a linear function of the strain rate. However, the stress could be a non-linear function of the strain rate in complex fluids such as polymer solutions,

$$\sigma = f(\mathbf{E}). \tag{2.97}$$

Since $\mathbf{E}_{ij}$ is a frame indifferent tensor, any tensor that can be written as

$$\sigma = \text{scalar} \times \mathbf{E} \tag{2.98}$$

satisfies the conditions of frame indifference. There are three frame indifferent scalars that can be constructed from the tensor $\nabla \mathbf{u}$:

$$I_1 = \nabla.\mathbf{u} \quad I_2 = \mathbf{E}{:}\mathbf{E} \quad I_3 = \text{Det}(\mathbf{E}). \tag{2.99}$$

Therefore, any constitutive relation that can be written in the form

$$\sigma = \eta(I_1, I_2, I_3)\mathbf{E} \tag{2.100}$$

would satisfy the requirements of material frame indifference. Of these frame-indifferent scalars, $I_1 = 0$ for an incompressible fluid.

The constitutive equation for the stress tensor can be inserted into the momentum conservation equation to obtain

$$\partial_t \rho + \partial_i (\rho v_i) = 0$$
$$\rho \partial_t v_i + \rho v_j \partial_j v_i = -\partial_i p + \eta \left(\partial_j^2 v_i - (2/3)\partial_i \partial_j v_j \right). \tag{2.101}$$

These are the Navier–Stokes mass and momentum conservation equations. For an incompressible fluid, where the density is a constant in both space and time, the Navier–Stokes equations have a particularly simple form:

$$\partial_i v_i = 0$$
$$\partial_t v_i + v_j \partial_j v_i = -\rho^{-1} \partial_i p + \nu \partial_j^2 v_i, \tag{2.102}$$

where $\nu = (\eta/\rho)$ is the kinematic viscosity.

2.8 Polymer Conformation

The most widely studied non-Newtonian fluids are polymer solutions and polymer melts. A polymer is a molecule with high molecular weight obtained by covalently bonding a large number of units or monomers, so that the linear length of the molecular is large compared to the molecular diameter. Polymers can be thought of as long flexible strings, and have many interesting properties because of their linear nature. In solutions and in melts, they are in a highly coiled state, as shown in Fig. 2.17. In solutions and melts, these springs are in a highly coiled state.

A useful physical picture is obtained if we consider these strings as undergoing a random walk [3]. The random walk proceeds in steps of equal length, called the 'Kuhn segment length' (which is a few times larger than the monomer size), with a random change in direction after each step. The average end-to-end distance of a random walk with N steps increases proportional to $\sqrt{N}$. Therefore, one would expect the linear size of a polymer molecule to increase proportional to $N^{1/2}$, where N is the number of monomers, if the molecule is accurately described by a random walk. However, in real solutions, there is an additional factor which comes into play, which is that the polymer molecule cannot cross itself. Due to this, the radius of gyration of a polymer molecule actually increases with a slightly higher power, $N^{0.58}$. If we assume a simple random walk model, then the average end-to-end distance of a polymer molecule increases proportional to $N^{1/2}$, where N is the number of Kuhn segments. The probability of finding a configuration with a particular end-to-end distance can be evaluated from entropic arguments, based on the number of configurations corresponding to this end-to-end distance. Based on simple arguments

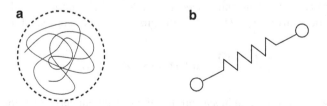

Fig. 2.17 (a) A polymer molecule and (b) the bead–spring representation of the polymer molecule

for a Gaussian chain, it can be inferred the probability that the end-to-end distance is x is proportional to $\exp(-x^2/(Nl^2))$, where l is the Kuhn segment length and N is the number of Kuhn segments. This is equivalent to a probability of the form $\exp(-E_x/kT)$, where the energy E_x for a molecule with end-to-end distance x is $E_x = (kT/Nl^2)x^2$. If we consider the factor $(2kT/Nl^2)$ as a spring constant k_n, then the energy penalty is of the form $E_x = (k_n x^2/2)$. This is just the stretching energy of a spring with spring-constant k_n. This leads to a still simpler bead–spring representation of a molecule, which is a linear spring between two beads which has a spring constant k_n, as shown in Fig. 2.17.

In the bead–spring model, the stress due to the polymers is expressed in terms of the second-order conformation tensor, and has a very simple form,

$$\mathbf{T}^{\mathrm{p}} = c_{\mathrm{p}} k_n \mathbf{xx} = c_{\mathrm{p}} k_n \mathbf{Q}, \tag{2.103}$$

where the term on the right contains the tensor product of the end-to-end displacement vector $\mathbf{x}$, and c_{p} is the polymer concentration. The tensor $\mathbf{Q} = \mathbf{xx}$ is referred to as the 'polymer conformation tensor', which basically provides information about the state of stretch of the polymers. Here, we have a situation where the local flow field affects the polymers by stretching and rotating them, and the state of stretch of the polymers provides an additional component of the stress which in turn affects the flow field. Therefore, in a fluid-dynamical description, it is necessary to have coupled equations for both the flow field and the conformation tensor of the polymers.

In the equation for the conformation tensor of the polymer, there are terms corresponding to the stretching and bending of the polymers due to the imposed flow, and there is a 'spring' force which tends to reduce the polymer conformation back to its equilibrium state. From the Gaussian approximation for the polymer end-to-end distance, the conformation tensor $\mathbf{Q}^{\mathrm{eq}}$ at equilibrium is an isotropic tensor with no preferred direction,

$$\mathbf{Q}^{\mathrm{eq}} = \frac{kT}{k_n}\mathbf{I}. \tag{2.104}$$

The simplest equation for the conformation tensor states that the polymer tends to relax back to its equilibrium conformation with a time constant,

$$\frac{\mathscr{D}\mathbf{Q}}{\mathscr{D}t} = -\frac{\mathbf{Q} - \mathbf{Q}^{\mathrm{eq}}}{\tau}. \tag{2.105}$$

There are two new terms introduced above, the relaxation time τ and the 'upper convected' derivative $(\mathscr{D}/\mathscr{D}t)$ on the left. First, the upper convected derivative is defined as

$$\frac{\mathscr{D}\mathbf{Q}}{\mathscr{D}t} = \frac{\partial\mathbf{Q}}{\partial t} + \mathbf{u}.\nabla\mathbf{Q} - \mathbf{Q}.\nabla\mathbf{u} - (\nabla\mathbf{u})^{\mathrm{T}}.\mathbf{Q}. \tag{2.106}$$

The first two terms on the right side are just the substantial derivatives, but the 'upper convected' derivative contains two additional terms. In our earlier discussion, we noted that the substantial derivative is a derivative in a reference frame moving with the local fluid mean velocity. In a similar manner, the 'upper convected' derivative also contains terms due to the local rotation of fluid elements at a point. Since the rotation of the fluid elements will tend to rotate any bead–spring dumbbells immersed in the fluid, this rotation is included in the substantial derivative. Therefore, the upper convected derivative is a derivative in a reference frame moving and rotating with the local fluid element. On the right side of (2.106), the time constant $\tau = (\xi/k_n)$ is the ratio of the spring constant and a 'friction coefficient' ξ, which represents the force exerted on the beads due to the fluid flow past the beads. If the fluid is flowing with a velocity $\mathbf{v}$ past the beads, the force exerted can be written as $\mathbf{F} = -\xi\mathbf{v}$ in the linear approximation. The ratio of this frictional force and the spring constant is the time constant.

Once the value of $\mathbf{Q}$ is known, then (2.103) can be used to evaluate the polymer stress, which is then inserted into the momentum conservation equation of the fluid,

$$\rho\left(\frac{\partial u}{\partial t} + \mathbf{u}.\nabla\mathbf{u}\right) = -\nabla p + \eta\nabla^2\mathbf{u} + \nabla.\mathbf{T}^p. \tag{2.107}$$

The polymer stress is usually written in an alternate, but completely equivalent, form to (2.103),

$$\mathbf{T}^p = \frac{\eta_p k_n}{\tau k T}(\mathbf{Q} - \mathbf{Q}^{eq}), \tag{2.108}$$

where $\eta_p = (c_p k T \tau)$ is the polymer viscosity, and the only difference between the (2.108) and (2.103) is that we have removed the isotropic part proportional to $\mathbf{Q}^{eq}$. (2.107) and (2.108) constitute the upper convected Maxwell model for a polymeric fluid.

References

1. Branes HA, Hutton JF, Walters K (1989) An introduction to rheology, Elsevier, Amsterdam
2. Bird RB, Stewart WE, Lightfoot EN (2001) Transport phenomena, Wiley, NY
3. de Gennes P-G (1979) Scaling concepts in polymer physics, Cornell University Press, NY

Chapter 3
Mechanics of Liquid Mixtures

Kumbakonam Ramamani Rajagopal

Abstract A brief introduction is provided for modeling the response of mixtures under the assumption that the constituents can be modeled as a continuum and that the constituents co-occupy the region of the mixture in a homogenized sense. The constituents of the mixture can undergo chemical reactions and there can be interconversion between the constituents. Balance laws are provided for the constituents of the mixture that allow for the chemical reactions as well as the numerous other interaction mechanisms between the constituents. After developing the general framework, the theory will be used to develop a model for the flow of a mixture of fluids.

3.1 Introduction

According to the Oxford English Dictionary [61], a mixture is defined as "...product of mixing a complex aggregate (material or immaterial) composed of various ingredients or constituent parts mixed together" and "...mixed state or condition; co-existence of different ingredients or different groups or classes of things mutually diffused through each other." Trying to get an understanding of a mixture as "a product of mixing" or being in a "mixed state" is tantamount to going around circles, but that is the usual result of using a dictionary for edification, one that requires an innate intuitive understanding of a set of primitive ideas and concepts, and we shall suppose the reader has such an understanding as to what constitutes a mixture.

Practically every substance one comes across is a mixture. To extract a pure substance entails prohibitive costs and even at best what one has is a nearly pure substance. Thus, one often deals with mixtures of various degrees of mixed constituents, and this article pertains to bodies wherein there is a sufficient amount

K.R. Rajagopal (✉)
Department of Mechanical Engineering, Texas A&M University, College Station,
Texas 77845, USA
e-mail: krajagopal@tamu.edu

J.M. Krishnan et al. (eds.), *Rheology of Complex Fluids*,
DOI 10.1007/978-1-4419-6494-6_3, © Springer Science+Business Media, LLC 2010

of each constituent being present in that any arbitrary small chunk of the substance contains a sufficiently large number of molecules of each constituent so that each of them in a "homogenized" sense constitutes a single continuum. Numerous bodies that have relevance to biophysics, geophysics, astrophysics, etc., such as air, water, gaseous mixtures and plasmas that surround the stars, blood, tissues, metallic alloys, etc., are mixtures.

The various constituents that comprise a mixture can usually diffuse through one another. This relative motion between the constituents can take place at the macroscale or at the atomic scale. Even in solids, such relative motion between the constituents takes place at the atomic level and is usually referred to as "interdiffusion". There has been a great deal of work in this area and the reader is referred to the books by Barrer [8] and Seitz [56] and the numerous references therein.

This short review concerns the study of the motion of mixtures from a macroscopic point of view and discusses a mathematical framework that has been developed to study the thermomechanics of such mixtures.

The mathematical framework that will be discussed traces its origins to the seminal works of Fick [28] and Darcy [25]. These original studies resulted in equations that are referred to as "Fick's law" and "Darcy's law". While these equations have proven to be quite useful, to elevate such crude and ad hoc approximations to the status of "laws" is bordering on the absurd. They can be shown to be very crude approximations of the balance of linear momentum under a plethora of assumptions and approximations (see Munaf et al. [43] for a detailed discussion of the status of "Darcy's law" within the context of the theory of mixtures). For instance, the early work of Darcy was generalized subsequently by Brinkman [10, 11] and Biot [12–14].

As we shall consider the thermomechanics of mixtures of fluids we shall document some of the early studies concerning the diffusion of one fluid through another. Significant contributions to this area have been made by Hadamard [31,32], Rybczynski [53] and Boussinesq [18]. Many models have been proposed, most of them ad hoc and some from certain physical considerations and statistical assumptions that lead to empirical formulae for emulsions (see Hatschek [34], Einstein [26, 27], Saltzer and Shultz [54], Taylor [60], Oldroyd [44], Barnea and Mirhazi [5], Tamura and Kurata [59], Al-Sharif et al. [2]) and particle suspensions (see Batchelor and Green [7], Brinkman [15], Simha [57], Quemada [67], Jeffery [36], Burgers [22], Massoudi [40], Rajagopal [45], Mooney [42], Brenner [16] and Brodnyan [17]).

There is no attempt here to provide an exhaustive documentation of the literature concerning the thermodynamics of mixtures. The interested reader will find numerous references in the book by Rajagopal and Tao [64] and the recent review by Rajagopal [47] and the papers cited therein. With regard to the theory of mixtures that will form the basis for the study of thermomechanics of mixtures studied herein, the reader is referred to the papers by Truesdell [63–65], the several appendices in his book on rational thermodynamics [66], the review articles by Bowen [21], Atkin and Craine [4], Bedford and Drumheller [9], and the books by Samohyl [55], and by Rajagopal and Tao [46].

3.2 Kinematics and the Basic Balance Laws

The fundamental assumption of the Theory of Mixtures is that at each point in the domain occupied by the mixture, there exists, in a homogenized sense, a particle belonging to each constituent of the mixture. Thus, one can visualize the mixtures as n homogenized single continua placed atop[1] one another with the ability of the particles belonging to each constituent to move with respect to the other constituents. We denote by $\mathbf{X}^i$, a typical particle belonging to a reference configuration $\kappa^i_R(\boldsymbol{B})$ and let x denote its position in the current configuration of the mixture. By the motion of the ith constituent, we mean a sufficiently smooth one to one mapping[2] $\chi^i(\mathbf{X}^i, t)$ such that

$$\mathbf{x}_i = \chi^i(\mathbf{X}^i, t), i = 1, 2, ...N, \tag{3.1}$$

where N denotes the number of constituents. With each constituent, we associate a density ρ^i. We define the total density ρ of the mixture through

$$\rho = \sum_{i=1}^{N} \rho^i. \tag{3.2}$$

The displacement of the particle $\mathbf{X}^i$ is defined through

$$\mathbf{u}^i = \mathbf{x} - \mathbf{X}^i, \tag{3.3}$$

and the velocity of the particle $\mathbf{X}^i$ is defined through

$$\mathbf{v}^i = \frac{\partial \chi^i}{\partial t}. \tag{3.4}$$

We can define an averaged mixture velocity v through

$$\mathbf{v} = \frac{1}{\rho} \sum_{i=1}^{N} \rho^i \mathbf{v}^i. \tag{3.5}$$

---

[1] The word "*atop*" is inappropriate but provides some basis for understanding the co-occupancy of the constituents. Each of the particles simultaneously co-exist at the same point in the mixture. Of course, there are serious physical problems with such an assumption, but this is far less worrisome from a philosophical stand point of existence than the notion of a "*point*", an entity with no dimension! We seem to have no problem whatsoever with coming to grips with this unfathomable concept. Since each particle belonging to a constituent has no dimension allowing them to co-occupy should not lead to any unsurmountable difficulty.

[2] Unless a summation symbol is explicitly used, repeated indices do not mean summation over that index.

However, it is imperative to recognize that this mixture velocity is quite arbitrary and we could have chosen to define it differently. Thus, physical constraints concerning the mixture have to be interpreted quite carefully. For instance, a mathematical constraint that a mixture velocity $\mathbf{v}$ satisfy div $\mathbf{v} = 0$, has different implications depending on how $\mathbf{v}$ is defined in terms of the quantities associated with the constituents. It is important to recognize that any property ϕ^i associated with the ith constituent can be expressed in terms of $\mathbf{X}^i$ and time t or $\mathbf{x}$ and time t. Thus ϕ can be expressed as

$$\phi^i = \phi^i(\mathbf{X}^i, t) = \hat{\phi}^i(\mathbf{x}, t). \tag{3.6}$$

Since the mapping χ^i is assumed to be invertible, we can now define the derivative

$$\nabla_{\mathbf{X}^i} \phi^i = \frac{\partial \phi^i}{\partial \mathbf{X}^i}, \tag{3.7}$$

$$\frac{D^i \phi^i}{Dt} = \frac{\partial \phi^i}{\partial t}, \tag{3.8}$$

$$\nabla_{\mathbf{x}^i} \phi^i = \frac{\partial \hat{\phi}^i}{\partial \mathbf{x}^i}, \tag{3.9}$$

$$\frac{\partial \phi^i}{\partial t} = \frac{\partial \hat{\phi}^i}{\partial t}. \tag{3.10}$$

One can also define a material time derivative associated with the mixture through

$$\frac{D\phi^i}{Dt} = \frac{\partial \phi^i}{\partial t} + \nabla_{\mathbf{x}} \hat{\phi} \mathbf{v}. \tag{3.11}$$

Once again it is important to recognize that these derivatives associated with the mixture depend on how the properties associated with the mixture are defined. We shall not define the deformation gradients, strains, etc., associated with the constituents as we are interested in mixtures of fluids. The velocity gradient $\mathbf{L}^i$ and its symmetric part and skew parts $\mathbf{D}^i$ and $\mathbf{W}^i$ are defined through:

$$\mathbf{D}^i = \frac{1}{2}\left(\mathbf{L}^i + (\mathbf{L}^i)^{\mathsf{T}}\right), \tag{3.12}$$

and

$$\mathbf{W}^i = \frac{1}{2}\left(\mathbf{L}^i - (\mathbf{L}^i)^{\mathsf{T}}\right). \tag{3.13}$$

We define the Rivlin-Ericksen tensors $\mathbf{A}_n^i$ through

$$\mathbf{A}_1^i = 2\mathbf{D}_1, \tag{3.14}$$

and

$$\mathbf{A}_n^i = \frac{D^i}{Dt}\mathbf{A}_{n-1}^i + \mathbf{A}_{n-1}^i\mathbf{L}^i + (\mathbf{L}^i)^T\mathbf{A}_{n-1}^i \tag{3.15}$$

The above kinematical definitions suffice for our discussions. We now proceed to record the balance equations.

3.3 Balance of Mass

The integral form of the balance of mass is given by

$$\frac{d}{dt}\int_{\wp_t}\rho^i\, dv + \int_{\partial\wp_t}\rho^i\mathbf{v}^i\cdot\mathbf{n}\, da = \int_{\wp_t} m^i\, dv,\ \forall\wp_t \subseteq \kappa\,(B)\,,\ i = 1,\dots, N. \tag{3.16}$$

In the above equation m^i is the mass of the ith constituent that is produced due to chemical reactions amongst the constituents, and $\mathbf{n}$ is the unit outward normal. The statement implies that the time rate of change of mass of the ith constituent in the part $\wp_t$ of the mixture equals the mass of the ith constituent that is fluxed into the part $\wp_t$ through the boundary, and the mass produced within the part $\wp_t$. When the constituents are non-reacting, $m^i \equiv 0\ \forall i = 1, 2, \dots\dots.N$.

The local form of balance of mass for the ith constituent is given by

$$\frac{\partial\rho^i}{\partial t} + \operatorname{div}\left(\rho^i\mathbf{v}_i\right) = m^i, \tag{3.17}$$

where m^i denotes the mass supply for the ith constituent and the divergence operation is with respect to $\mathbf{x}$. In general, when one allows for interconversion between constituents such as in a chemically reacting mixture, $m^i \neq 0$. However, the balance of mass for all the constituents implies that $\displaystyle\sum_{i=1}^{N} m^i = 0$. When one sums (3.17) over the constituents, and uses (3.2) and (3.5), one obtains

$$\frac{\partial\rho}{\partial t} = \operatorname{div}\left(\rho\mathbf{v}\right) = 0. \tag{3.18}$$

In fact, it is the choice for $\mathbf{v}$ that is critical in obtaining the above form for the balance of mass for the total mixture equation. A different choice for $\mathbf{v}$ would have led to an equation that does not look like that for a single constituent. This, however, in no way means that it is inappropriate or incorrect. We merely do not obtain a form with which we are familiar and comfortable.

An interesting and useful discussion of stoichiometric issues can be found in the review article by Bowen [21] and the book by Samohyl [55].

3.4   Balance of Linear Momentum

Before recording the balance of linear momentum, we introduce a partial traction vector $\mathbf{t}^i$ with the ith constituent. The partial stress $\mathbf{T}^i$ associated with $\mathbf{t}^i$ for the ith constituent is given by

$$\left(\mathbf{T}^i\right)^T \mathbf{n} = \mathbf{t}^i, \quad i = 1, 2, \ldots\ldots, N, \tag{3.19}$$

where $\mathbf{n}$ is the unit outward normal at $\mathbf{x}$ to the surface on which $\mathbf{t}^i$ acts.

The integral form of the balance of linear momentum is given by

$$\frac{d}{dt} \int_{\wp_t} \rho^i \, dv + \int_{\partial \wp_t} \rho^i \mathbf{v}^i \cdot \mathbf{n} \, da = \int_{\partial \wp_t} \mathbf{t}^i \, da + \int_{\wp_t} \rho^i \mathbf{b}_e{}^i \, dv + \int_{\wp_t} \mathbf{I}^i \, dv$$

$$+ \int_{\wp_t} m^i \mathbf{v}^i \, dv, \ \forall \wp_t \subseteq \kappa\left(B\right), \ i = 1, \ldots, N. \tag{3.20}$$

In the above equation $\mathbf{b}^i$ is the external body force acting on the ith constituent, and $\mathbf{I}^i$ is the force acting on the ith constituent due to its interactions (that stems from its co-occupancy with other constituents at each point in the mixture) with the other components of the mixture. The above equation states that the time rate of change of linear momentum of any part of the mixture corresponds to momentum flux into the part and the forces due to the traction and body forces acting on the ith constituent, and the momentum that is produced that is associated with the mass production of the ith constituent, m^i.

The local form of the balance of linear momentum takes the form

$$\rho^i \frac{D\mathbf{v}^i}{Dt} = \operatorname{div}\left(\mathbf{T}^i\right)^T + \rho^i \mathbf{b}_e^i + \mathbf{I}^i, \tag{3.21}$$

where $\mathbf{b}_e^i$ is the external body force acting on the ith constituent and $\mathbf{I}^i$ denotes the forces due to the interaction of all the other constituents, on the ith constituent, in virtue of their being required to co-occupy each point of the mixture in a homogenized sense.

The precise meaning of the notion of a partial stress is a matter of some debate; one possible explanation is that it represents the stress borne by the ith constituent, in the homogenized state.

The interaction terms $\mathbf{I}^i$ are such that when they are summed over i, as a consequence of Newton's third law, they have to add up to zero, i.e,

$$\sum_{i=1}^{N}(\mathbf{I}^i + m^i \cdot v^i) = \mathbf{0}. \tag{3.22}$$

Numerous interaction forces come into play and many of the models that have been developed to describe the mechanics of swelling and of the flow through porous solids assume some specific form for the interactions. For instance, some of the interaction mechanisms that come into play are Drag, Virtual Mass Effect, Lift, Magnus Effect, Basset Forces, Faxen Forces, and effects due to temperature and density gradients. We shall not get into the discussion of the various effects here but refer the reader to Johnson et al. [35] where these are discussed in detail. We will just mention that understanding and modeling such interactions have occupied the interest of the best of scientists (see Thomson and Tait [62], Kirchoff [37], Lamb [38], Craig [23], Greenhill [30], Stokes [58], Rayleigh [52], Basset [6] and many others). References concerning the forces acting on particles due to a fluid can be found in Happel and Breuner [33].

It is worth mentioning that great care has to be exercised in providing a mathematical expression for the interaction term as it has to satisfy certain basic invariance requirements, such as Galilean invariance or the stronger requirement of frame-indifference. Relative acceleration between the constituents does not satisfy such invariance requirements and has been used a great deal in the literature involving fluids containing particulate media (an example of such an error can be found in the work of Anderson and Jackson [3] that is referred to copiously in the literature pertinent to fluidized body).

3.5 Balance of Angular Momentum

The balance of angular momentum (moment of momentum) in the integral form is given by

$$\frac{\mathrm{d}}{\mathrm{d}t} \int_{\wp_t} (\mathbf{x} - \mathbf{y}) \times \rho^i \mathbf{v}^i \mathrm{d}v + \int_{\partial \wp_t} (\mathbf{x} - \mathbf{y}) \times \rho^i \mathbf{v}^i (\mathbf{v}^i . \mathbf{n}) \mathrm{d}n$$

$$= \int_{\partial \wp_t} (\mathbf{x} - \mathbf{y}) \times \mathbf{t}^i \mathrm{d}a + \int_{\wp_t} (\mathbf{x} - \mathbf{y}) \times \rho^i \mathbf{b}^i \mathrm{d}v + \int_{\wp_t} (\mathbf{x} - \mathbf{y})$$

$$\times (\mathbf{I}^i + m^i \mathbf{v}^i) \mathrm{d}v, \quad \forall \wp_t \subseteq \kappa(B), \ i = 1, ..., N, \tag{3.23}$$

where $\mathbf{y}$ is any point in space. In the case of a single constituent, the balance law of angular momentum states that the Cauchy stress tensor is symmetric in the absence of body couples. However, in the case of mixtures, one can show that

$$\mathbf{T}^i - \left(\mathbf{T}^i\right)^{\mathrm{T}} = \boldsymbol{\varepsilon} \mathbf{M}^i \tag{3.24}$$

where $\boldsymbol{\varepsilon}$ is the alternator tensor and $\mathbf{M}^i$ is the angular momentum supplied to the ith constituent. We note that if $\sum \boldsymbol{\varepsilon} \mathbf{M}^i = 0$ then the total stress tensor is symmetric.

3.6 Balance of Energy

The integral form of the balance of energy is given by

$$
\frac{d}{dt} \int_{\wp_t} \rho^i \left(e^i + \frac{1}{2} |\mathbf{v}^i|^{(2)} \right) dv + \int_{\partial \wp_t} \rho^i \left(e^i + \frac{1}{2} |\mathbf{v}^i|^{(2)} \right) (\mathbf{v}^i \cdot \mathbf{n}) \, da
$$

$$
= \int_{\partial \wp_t} \mathbf{t}^i \cdot \mathbf{v}^i \, da + \int_{\wp_t} \rho^i \mathbf{b}^i \cdot \mathbf{v}^i \, dv + \int_{\wp_t} \mathbf{I}^i \cdot \mathbf{v}^i \, dv - \int_{\wp_t} \mathbf{q}^i \cdot \mathbf{n}^i \, da
$$

$$
+ \int_{\wp_t} \rho^i r^i \, dv + \int_{\wp_t} m^i e^i \, dv + \int_{\wp_t} e_s^i \, dv, \quad \forall \wp_t \subseteq \kappa (B), \quad i = 1, ..., N \quad (3.25)
$$

In the above equation, e_s^i is the energy supplied to the ith constituent by the other constituents. Also, the above equation states that the time rate of change of the energy of a constituent in a part of the mixture equals the energy flux of the ith constituent to that part plus the rate at which work is done by traction, body force and interaction forces associated with the ith constituent, the heat flux and radiation heating as well as the energy associated with the mass that has been created and the energy supplied to the ith constituent.

Let $e^i, \mathbf{q}^i, r^i$ and e_s^i denote the specific internal energy, heat flux vector, the radiant heating and the specific internal energy supply to the ith constituent. Then the balance of energy for the ith constituent can be expressed as

$$
\frac{\partial}{\partial t} \left(\rho^i e^i \right) + \text{div} \left(\rho^i e^i \mathbf{v}^i \right) = -\text{div}\, \mathbf{q}^i + \mathbf{T}^i \cdot \mathbf{L}^i + \rho^i r^i + e_s^i + m^i e^i \quad (3.26)
$$

On summing over the constituent, the above equation leads to

$$
\rho \frac{de}{dt} = \sum_{i=1}^{N} \left[\mathbf{T}^i \cdot \mathbf{L}^i - \mathbf{m}^i \cdot \mathbf{v}^i - \frac{1}{2} m^i \mathbf{v}^i \cdot \mathbf{v}^i \right] - \text{div}\, \mathbf{h} + \rho r, \quad (3.27)
$$

where,

$$
e = \frac{1}{\rho} \sum_{i=1}^{N} \rho^i e^i, \quad \mathbf{q} = \sum_{i=1}^{N} \mathbf{q}^i, \quad \mathbf{T} = \frac{1}{\rho} \sum_{i=1}^{N} \rho^i r^i, \quad (3.28)
$$

and

$$
\mathbf{h} = \mathbf{q} + \frac{1}{\rho} \sum_{i=1}^{N} \rho^i e^i \mathbf{v}^i. \quad (3.29)
$$

3.7 Second Law of Thermodynamics

While dealing with a mixture of two or more fluids one can start by assuming that the quantities that are to be specified quantitatively depend on

$$
\mathbf{v}, \ \mathbf{F}^i, \ \frac{\partial \mathbf{F}^i}{\partial \mathbf{X}^i}, \ \mathbf{L}^i, \ \rho^i, \ \frac{\partial \rho^i}{\partial \mathbf{x}^i}, \ \theta, \text{ and } \frac{\partial \theta}{\partial \mathbf{x}}.
$$

If the constituents are inhomogeneous, then one has to also assume that they depend on $\mathbf{X}^i$. One could also include the dependence on time t to incorporate effects, such as aging and denaturation of organic liquids. Models where the dependence of gradient of density are ignored are called simple fluid mixtures (see Samohyl [21] for a discussion of the relevant issues).

Even within the context of a single constituent continuum there are several questions concerning the precise form that second law ought to take; for instance, whether it should be enforced as a local inequality or as a global inequality, whether the Clausius–Duhem inequality is the appropriate form for the second law, etc. In the case of a mixture, there are additional problems, namely whether one can enforce it to hold for each constituent or only to the mixture as a whole. We shall not get into a discussion of these issues here but refer the reader to the book of Rajagopal and Tao [46], where some of the relevant issues are discussed.

More recently, Rajagopal and co-workers, based on earlier work that can be traced back to ideas of Rayleigh and Maxwell, and articulated under certain special considerations by Ziegler and co-workers, have advocated enforcing a stronger requirement than the second law, namely that the rate of entropy production be maximized, and using such an approach Malek and Rajagopal [39] studied the thermodynamics of fluid mixtures. In view of the above comments, we shall not postulate an integral statement of the second law for each constituent for an arbitrary part, nor provide an integral law for the mixture as a whole. We shall assume that we can write down the local form of the second law, a rather strong assumption, but then require it of only the mixture as a whole and not the individual constituents, a much weaker assumption. Here, we shall restrict ourselves to the local interpretation of the Clausius–Duhem inequality for the mixture as an appropriate form of second law.

Let η^i and θ^i denote the specific entropy and the absolute temperature associated with the i th constituent, respectively. We shall require that

$$-\rho \left(\dot{\psi} + \eta \dot{\theta} \right) + \left\{ \sum_{i=1}^{N} \operatorname{tr} \left[\mathbf{T}^i - \rho^i \left(\psi^i - \psi \right) \mathbf{1} \right] \mathbf{L}^i \right\} - \frac{1}{2} m^i \mathbf{v}^i \cdot \mathbf{v}^i$$

$$- \left[\mathbf{m}^i + \nabla \left(\rho^i \left(\psi^i - \psi \right) \right) \right] \cdot \mathbf{v}^i - \frac{\left[\mathbf{q}^i + \rho^i \theta \eta^i \right] \cdot \nabla \theta}{\theta} \geq 0, \qquad (3.30)$$

where

$$\eta = \frac{\displaystyle\sum_{i=1}^{N} \rho^i \eta^i}{\rho}, \quad \psi = \frac{\displaystyle\sum_{i=1}^{N} \rho^i \psi^i}{\rho}. \qquad (3.31)$$

Here, ψ^i is the specific Helmholtz potential defined through

$$\psi^i = e^i - \theta^i \eta^i \qquad (3.32)$$

and we have assumed that all the constituents have associated with the same temperature, i.e, $\theta^i = \theta, \forall i$. This is not always the case; a simple example is that of a mixture of two liquids with very great temperature difference. The assumption of the same temperature for both the constituents implies that the two liquids will have to attain the same temperature instantaneously. We shall, however, assume that the two bodies have the same temperature.

3.8   Volume Additivity Constraint

We next discuss a constraint which seems reasonable when dealing with a mixture of liquids. Such a constraint is not reasonable when dealing with a mixture, one of whose constituents is a gas. When dealing with a single continuum, the constraint that the continuum is incompressible leads to the requirement that all its motion has to be isochoric. In dealing with mixtures, when we have two constituents which are in their pure states incompressible, we expect the total volume of the mixture to be the sum of the individual volumes in all motions. This interpretation, however, is fraught with a lot of dangers. If the constituents are incompressible in their pure states, it would not be possible to mix them together so that in the mixed state their volumes are additive as at each point in the mixture we need a point belonging to the constituent. But here we are really referring to a particle belonging to the homogenized continuum that acts as a reference for the motions of the constituent. Thus, in effect we have two reference states and the mapping from the first to the second is not volume preserving. Further subtle issues come into play but we shall not discuss them here, but we shall merely follow the volume additivity constraint developed by Mills [41].

$$\sum_{i=1}^{N} \frac{\rho^i}{\rho_R^i} = 1, \tag{3.33}$$

or equivalently

$$\sum_{\substack{i=1 \\ j=1 \\ i \neq j}}^{N} \frac{\rho^i}{\rho_R^i} \left(\mathrm{div}\, \mathbf{v}^i \right) + \nabla \left(\frac{\rho^i}{\rho_R^i} \right) \left(\mathbf{v}^i - \mathbf{v}^j \right) = 0. \tag{3.34}$$

3.9   Constitutive Relations

There have been numerous studies concerning the thermomechanics of the mixture of fluids (see Mills [41], Adkins [1], Green and Nagdhi [29], Craine et al. [24], Bowen [19], Williams and Sampaio [68], Rajagopal and Tao [46], Malek and Rajagopal [39] and references therein). Here, we shall introduce a very simple model just to illustrate the ideas involved. Let us restrict our discussion to a mixture of two fluids. It is possible that a mixture of two linearly viscous fluids (Navier–Stokes fluids) could be non-Newtonian and also a mixture of two non-Newtonian fluids could behave as a linearly viscous fluid. In general, the partial stresses associated with the fluids could depend on the density, the temperature and the velocity gradient of the fluids. In fact, it could depend on several other quantities, such as density gradient, Rivlin–Ericksen tensors associated with the constituent (or even the history of motion associated with each of the constituents), etc.

While dealing with a mixture of two or more fluids one can start by assuming that the quantities that are to be specified quantitatively depend on

$$\mathbf{v}, \mathbf{F}^i, \frac{\partial \mathbf{F}^i}{\partial \mathbf{X}^i}, \mathbf{L}^i, \rho^i, \frac{\partial \rho^i}{\partial x^i}, \theta, \quad \text{and} \quad \frac{\partial \theta}{\partial \mathbf{x}}.$$

If the constituents are inhomogeneous, then one has to also assume that they depend on $\mathbf{X}^i$. We shall assume a simple structure

$$\mathbf{T}^{(1)} = \mathbf{f}^{(1)}\left(\rho^{(1)}, \rho^{(2)}, \frac{\partial \rho^{(1)}}{\partial \mathbf{x}}, \frac{\partial \rho^{(2)}}{\partial \mathbf{x}}, \mathbf{D}^{(1)}, \mathbf{D}^{(2)}\right), \tag{3.35}$$

and

$$\mathbf{T}^{(2)} = \mathbf{f}^{(2)}\left(\rho^{(1)}, \rho^{(2)}, \frac{\partial \rho^{(1)}}{\partial \mathbf{x}}, \frac{\partial \rho^{(2)}}{\partial \mathbf{x}}, \mathbf{D}^{(1)}, \mathbf{D}^{(2)}\right). \tag{3.36}$$

One needs the density gradients if one needs models of sufficient generality. However, one can consider a much simpler class of models of the following type for a mixture of liquids:

$$\mathbf{T}^{(1)} = -p\frac{\rho^{(1)}}{\rho}\mathbf{1} + \mu_{11}\mathbf{D}^{(1)} + \mu_{12}\mathbf{D}^{(2)}, \tag{3.37}$$

$$\mathbf{T}^{(2)} = -p\frac{\rho^{(2)}}{\rho}\mathbf{1} + \mu_{21}\mathbf{D}^{(1)} + \mu_{22}\mathbf{D}^{(2)}, \tag{3.38}$$

and

$$\mathbf{I} = \mathbf{I}^{(1)} = -\mathbf{I}^{(2)} = \alpha\left(\mathbf{v}^{(1)} - \mathbf{v}^{(2)}\right). \tag{3.39}$$

Even the above structure for the equations leads to a very complicated problem as we would have to deal with coupled non-linear partial differential equations. What makes the task even more onerous is the fact that the prescription of boundary conditions is far from settled. For the problem under consideration, we could make the assumption concerning the velocity associated with each constituent. This is, however, quite ad hoc as all we know at best, in a real problem is the velocity of the mixture at the boundary and not what it is for each constituent. However, this is not such a great leap of faith from that which is taken in the case of a single constituent. The boundary conditions for the velocity or the stress are far from clear for a single component fluid. The usual no-slip boundary condition is just a mere assumption. In fact, to specify the condition at the boundary, one needs to know the nature of the constituents on either side of the boundary. The early masters such as Navier, Poisson, Girard, Stokes and others clearly understood this and tried to *derive* conditions that would obtain at the boundary. Unfortunately, this aspect with regard to boundary conditions has been forgotten by both applied researchers such as physicists and engineers who have raised the status of conditions such as the no-slip boundary condition to that of a principle, and the mathematicians who supply Dirichlet and Neumann conditions with gay abandon without worrying about the physics of the problem. The interested reader can find issues concerning boundary conditions and the resolution of the boundary conditions within the context of some boundary value problems within the context of mixture theory in the book by Rajagopal and Tao [46].

3.10   Constitutive Relations by Appealing to the Maximization of the Rate of Entropy Production

In the traditional approach in continuum thermodynamics, one assumes constitutive structures for the stress **T**, heat flux vector, entropy, internal energy, etc., and then uses the inequality (3.30), namely the second law of thermodynamics, to obtain restrictions on the form of the constitutive relations by requiring that the body can undergo arbitrary process. The problem with such an approach is that the constitutive assumption that we start with is not meant to hold in arbitrary process. Thus, it would be better to satisfy sufficient conditions under which we can ensure the inequality (3.30), rather than use arbitrary motions to obtain necessary and sufficient conditions. Moreover, it is possible that the requirement (3.30) is too weak in that too many candidates for constitutive relations might meet the inequality. Also to start with, we make constitutive assumptions for the stress, which is a tensor. Hyperelasticity provides an interesting example wherein, instead of making an assumption for the stress that is a tensor we make an assumption for the stored energy that is a scalar and derive the form for the stress. In dissipative materials that can also store energy, one can also assume constitutive structures on the stored energy and the rate of dissipation (in general the rate of entropy production) and obtain the constitutive relation for the stress.

Recently, a general thermodynamics framework has been put into place, wherein one can obtain constitutive relations for a disparate class of material response: viscoelasticity, inelasticity due to slip, twinning, solid to solid phase transition, solidification and melting, and also the response of mixtures, by requiring that all processes of the body meet the requirement that rate of entropy production is maximized. Ziegler [69, 70] and Wehrli and Ziegler [71] advocated the use of such an idea in plasticity, but within a totally different context than that used by Rajagopal and co-workers (see Rajagopal and Srinivasa [48] for a detailed discussion of the differences in the two approaches as well as some mathematical errors in the work of Ziegler). Also, Ziegler's approach will not allow one to obtain a large class of constitutive response relations (see Rajagopal and Srinivasa [49, 50]).

When a body produces entropy during a process, it invariably leads to a change in the "natural configuration" of the body, the "natural configuration" being the configuration to which the body would go to on the removal of external stimuli. The maximization of the rate of entropy production determines how the "natural configuration" would evolve. Such a framework is ideally suited to study the response of bodies whose "natural configurations" evolve during the process. The thermodynamic framework discussed in [20] and [68], uses the Helmholtz potential of the body that depends on kinematical quantities measured from a reference configuration and the natural configuration. Such an approach is particularly well suited to study the response of bodies which can be viewed as a class of response functions from an evolving set of natural configurations. However, it is also possible to use the idea of the maximization of the rate of dissipation function without using the fact that the natural configuration changes but to use a Gibbs potential that depends on the Cauchy stress, that is a quantity associated with the current configuration (see the recent paper by Rajagopal and Srinivasa [51] for a discussion of such an approach).

Recently, Malek and Rajagopal [39] used the above thermodynamic framework discussed in [20] and [68], to study the thermomechanics of a mixture of two liquids. We shall end this chapter with a discussion of this work. We express the second law (3.30) in a slightly different manner, namely

$$\frac{\partial}{\partial t}\left(\sum \rho^i \eta^i\right) + \text{div}\left(\sum \rho^i \eta^i \mathbf{v}^i\right) = \frac{r}{\theta} - \text{div}\left(\frac{\mathbf{q}}{\theta}\right) + \zeta, \zeta \geq 0, \qquad (3.40)$$

where ζ represents the rate of entropy production associated with the mixture as a whole. On introducing the Helmholtz potential $\psi := \varepsilon - \theta\eta$, it follows that

$$-\frac{\partial}{\partial t}\left(\sum \rho^i \psi^i\right) - \text{div}\left(\sum \rho^i \psi^i \mathbf{v}^i\right) + \sum \left(\mathbf{T}^i \cdot \nabla \mathbf{v}^i\right) + \mathbf{m}^{(1)} \cdot \left(\mathbf{v}^{(2)} - \mathbf{v}^{(1)}\right) = \xi. \tag{3.41}$$

In writing the above equation, we have assumed that we have a mixture of two liquids.

For simplicity of illustration, let us consider a purely mechanical problem, i.e., we will neglect thermal variables altogether. We shall also require that the Helmholtz potential for the mixture as a whole meets

$$\sum \left(\mathbf{T}^i \cdot \nabla \mathbf{v}^i\right) + \mathbf{m}^{(1)} \cdot \left(\mathbf{v}^{(2)} - \mathbf{v}^{(1)}\right) - \rho \frac{d\psi}{dt} = \xi. \tag{3.42}$$

The above equation implies that $\psi^{(1)}$ and $\psi^{(2)}$ have to be of a certain form. If one assumes $\psi = \psi^{(1)} = \psi^{(2)}$, the above condition will be satisfied. Let us assume that

$$\psi = \psi\left(\rho^{(1)}, \rho^{(2)}\right) = \psi(\rho) \tag{3.43}$$

i.e., we assume a very special form of the specific Helmholtz potential for the mixture, namely that it only depends on the total density. Let us assume a rate of dissipation of the form

$$\xi = 2\mu\left(\rho\right)\|\mathbf{D}\|^{(2)} + \lambda\left(\rho\right)\left(\mathrm{tr}\,\|\mathbf{D}\|^{(2)}\right) + \alpha\left(\rho\right)\left\|\mathbf{v}^{(2)} - \mathbf{v}^{(1)}\right\|^{(2)}, \tag{3.44}$$

where

$$\mathbf{D} = \frac{1}{2}\left[\mathbf{L} + \mathbf{L}^{\mathsf{T}}\right], \quad \mathbf{L} = \nabla\mathbf{v}$$
$$= \frac{\rho^{(1)}}{\rho}\mathbf{D}^{(1)} + \frac{\rho^{(2)}}{\rho}\mathbf{D}^{(2)} + \frac{1}{\rho^{(2)}}\left[\left(\mathbf{v}^{(2)} - \mathbf{v}^{(1)}\right) \otimes \left(\rho^{(1)}\nabla\rho^{(2)} - \rho^{(2)}\nabla\rho^{(1)}\right)\right] \tag{3.45}$$

Let us suppose the mixture meets the constraint

$$\mathrm{tr}\,\mathbf{D} = 0. \tag{3.46}$$

This constraint does not mean that the individual constituents can only undergo isochoric motion. Also, as one could define the velocity for a mixture in different ways, the constraint has to be interpreted within the context of how the mixture velocity is defined. It follows from (3.5) that (3.46) implies that

$$\mathrm{tr}\left[\nabla\left(\frac{1}{\rho}\left(\rho^{(1)}\mathbf{v}^{(1)} + \rho^{(2)}\mathbf{v}^{(2)}\right)\right)\right] = 0. \tag{3.47}$$

We notice that in addition to the usual dissipation in a viscous fluid we also have dissipation due to the drag that is a consequence of one fluid diffusing through the other.

For any tensor $\mathbf{A}$, let $\mathbf{A}^{\mathrm{d}}$ be defined through

$$\mathbf{A}^{\mathrm{d}} = \mathbf{A} - \frac{1}{3}\left(\mathrm{tr}\,\mathbf{A}\right)\mathbf{I}, \tag{3.48}$$

i.e., $\mathbf{A}^{\mathrm{d}}$ is the deviatoric part of $\mathbf{A}$. Then, it follows that

$$\xi = 2\mu\left(\rho\right)\|\mathbf{D}\|^{(2)} + \lambda(\rho)\left(\mathrm{tr}\,\|\mathbf{D}\|^{(2)}\right) + \alpha\left(\rho\right)\left\|\mathbf{v}^{(2)} - \mathbf{v}^{(1)}\right\|^{(2)}, \tag{3.49}$$

and we have written the rate of dissipation as the sum of that due to volume change and that due to the drag. A simple calculation leads to

$$\sum\left[\mathbf{T}^i\right]^{\mathrm{d}}\cdot\left[\mathbf{L}^i\right]^{\mathrm{d}} + \sum\left(\frac{1}{3}\mathrm{tr}\,\mathbf{T}^i\,\mathrm{div}\,\mathbf{v}^i\right) + \rho^{(2)}\acute{\psi}\left(\rho\right)\mathrm{div}\,\mathbf{v} + \mathbf{m}^{(1)}\cdot\left(\mathbf{v}^{(2)} - \mathbf{v}^{(1)}\right) = \xi. \tag{3.50}$$

By maximizing the rate of entropy production, subject to the above equation as a constraint, we can obtain constitutive relations for

$$\left[\mathbf{T}^{(1)}\right]^{\mathrm{d}}, \left[\mathbf{T}^{(2)}\right]^{\mathrm{d}}, \frac{1}{3}\mathrm{tr}\,\mathbf{T}^{(1)}, \frac{1}{3}\mathrm{tr}\,\mathbf{T}^{(2)}, \text{ and } \mathbf{m}^{(1)}$$

by maximizing with respect to

$$\left[\mathbf{L}^{i}\right]^{\mathrm{d}}, \left[\mathbf{L}^{i}\right]^{\mathrm{d}}, \operatorname{div}\mathbf{v}^{(1)}, \operatorname{div}\mathbf{v}^{(2)} \text{ and } \mathbf{v}^{(2)} - \mathbf{v}^{(1)}.$$

We shall not provide the details of the calculations, the interested reader can find them in Malek and Rajagopal [39]. Before providing the consequences of the maximization of the rate of dissipation, subject to (3.46) as a constraint, it is useful to express (3.50) in the following form (see Malek and Rajagopal [39]):

$$\sum \left[\mathbf{T}^{i}\right]^{\mathrm{d}} \cdot \left[\mathbf{L}^{i}\right]^{\mathrm{d}} + \sum \left(\frac{1}{3}\mathrm{tr}\,\mathbf{T}^{i} + \rho^{i}\rho\dot{\psi}\left(\rho\right)\right)\operatorname{div}\mathbf{v}^{i}$$

$$+ \left(\mathbf{m}^{(1)} + \dot{\psi}\left(\rho\right)\boldsymbol{\varUpsilon}^{12}\right) \cdot \left(\mathbf{v}^{(2)} - \mathbf{v}^{(1)}\right) = \xi, \qquad (3.51)$$

where

$$\boldsymbol{\varUpsilon}^{12} := \rho^{(1)}\nabla\rho^{(2)} - \rho^{(2)}\nabla\rho^{(1)}. \qquad (3.52)$$

We shall enforce (3.51) as a constraint when we carry out the maximization. It then follows from the constrained maximization that

$$\mathbf{T}^{(1)} = \frac{1}{3}\left(\mathrm{tr}\,\mathbf{T}^{(1)}\right)\mathbf{I} + \left[\mathbf{T}^{(1)}\right]^{\mathrm{d}}$$

$$= -\rho^{(1)}\rho\dot{\psi}\left(\rho\right)\mathbf{I} + \left(\frac{2\mu\left(\rho\right)}{3} + \lambda(\rho)\right)\frac{\rho^{(1)}}{\rho}\left(\operatorname{div}\mathbf{v}\right)\mathbf{I} + 2\mu\frac{\rho^{(1)}}{\rho}\mathbf{D}^{d}, \quad (3.53)$$

$$\mathbf{T}^{(2)} = \frac{1}{3}\left(\mathrm{tr}\,\mathbf{T}^{(2)}\right)\mathbf{I} + \left[\mathbf{T}^{(2)}\right]^{\mathrm{d}}$$

$$= -\rho^{(2)}\rho\dot{\psi}\left(\rho\right)\mathbf{I} + \left(\frac{2\mu\left(\rho\right)}{3} + \lambda(\rho)\right)\frac{\rho^{(2)}}{\rho}\left(\operatorname{div}\mathbf{v}\right)\mathbf{I} + 2\mu\frac{\rho^{(2)}}{\rho}\mathbf{D}^{d}, \quad (3.54)$$

$$\mathbf{m}^{(1)} = \left(\frac{1}{\rho^{2}}\left(\frac{2\mu\left(\rho\right)}{3} + \lambda\left(\rho\right)\right)\operatorname{div}\mathbf{v} - \dot{\psi}\left(\rho\right)\right)\left(\rho^{(1)}\nabla\rho^{(2)} - \rho^{(2)}\nabla\rho^{(1)}\right)$$

$$+ \frac{2\mu}{\rho^{2}}\mathbf{D}^{d}\left(\rho^{(1)}\nabla\rho^{(2)} - \rho^{(2)}\nabla\rho^{(1)}\right) + \alpha\left(\rho\right)\left(\mathbf{v}^{(2)} - \mathbf{v}^{(1)}\right) \qquad (3.55)$$

We note that $\mathbf{T}^{(1)}$ and $\mathbf{T}^{(2)}$ are symmetric in virtue of our choice for the rate of dissipation. In general, they need not be symmetric. We could maximize the rate of dissipation subject to both (3.46) and (3.51) as constraints. In this case, we will find that

$$\mathbf{T}^{(1)} = \frac{1}{3}\left(\operatorname{tr}\mathbf{T}^{(1)}\right)\mathbf{I} + \left[\mathbf{T}^{(1)}\right]^{\mathrm{d}} = (-p\mathbf{I} + 2\mu\,(\rho)\,\mathbf{D})\,\frac{\rho^{(1)}}{\rho}, \qquad (3.56)$$

$$\mathbf{T}^{(2)} = \frac{1}{3}\left(\operatorname{tr}\mathbf{T}^{(2)}\right)\mathbf{I} + \left[\mathbf{T}^{(2)}\right]^{\mathrm{d}} = (-p\mathbf{I} + 2\mu\,(\rho)\,\mathbf{D})\,\frac{\rho^{(2)}}{\rho}, \qquad (3.57)$$

and

$$\mathbf{m}^{(1)} = (-p\mathbf{I} + 2\mu\,(\rho)\,\mathbf{D})\,\frac{\rho^{(1)}\nabla\rho^{(2)} - \rho^{(2)}\nabla\rho^{(1)}}{\rho^2} + \alpha\,(\rho)\left(\mathbf{v}^{(2)} - \mathbf{v}^{(1)}\right). \quad (3.58)$$

We see that in both the cases, we obtain expressions for the partial stresses, which when added up to obtain the total stresses in the mixture has the form for the classical compressible and incompressible fluid models, respectively.

References

1. Adkins JE (1963) Nonlinear diffusion I. Diffusion and flow of mixtures of fluids. Philos Trans R Soc Lond A 225:607–633
2. Al-Sharif A, Chamniprasart K, Rajagopal KR, Szeri AZ (1993) Lubrication with binary mixtures: Liquid–liquid emulsion. J Tribol 115(1):46–55
3. Anderson TB, Jackson R (1968) A fluid mechanical description of fluidized beds: Stability of the state of uniform fluidization. Ind Eng Chem Fund 7:12-21
4. Atkin RJ, Craine RE (1976) Continuum theories of mixtures: Applications. J Inst Math Appl 17:153–207
5. Barnea E, Mizrahi J (1976) On the effective viscosity of liquid–liquid dispersions. Ind Eng Chem Fund 15:120–125
6. Basset AB (1888) Treatise on hydrodynamics. Deighton Bell, Cambridge
7. Batchelor G, Green JT (1972) The determination of the bulk stress in a suspension of spherical particles to order $C^{(2)}$. J Fluid Mech 56:401–427
8. Barrer RM (1941) Diffusion in and through solids. Cambridge University Press, London
9. Bedford A, Drumheller DS (1983) Recent advances: Theories of immiscible and structured mixtures. Int J Eng Sci 21:863–960
10. Brinkman HC (1947) On the permeability of media consisting of closely packed porous particles. Appl Sci Res A 1:81–86
11. Brinkman HC (1947) The calculation of the viscous force exerted by a flowing fluid on a dense swarm of particles. Appl Sci Res A 1:27–34
12. Biot MA (1934) Theory of elastic waves in a fluid-saturated porous solid, I. Low frequency range. J Acoust Soc Am 28:168–178 DOI:10.1121/1.1908239
13. Biot MA (1934) Theory of elastic waves in a fluid-saturated porous solid, II. High frequency range. J Acoust Soc Am 28:179–191 DOI:10.1121/1.1908241
14. Biot MA (1962) Mechanics of deformation and acoustic propagation in porous media. J Appl Phys 33:1482–1498
15. Brinkman HC (1952) The viscosity of concentrated suspensions and solutions. J Chem Phys 20:571
16. Brenner H (1958) Dissipation of energy due to solid particles suspended in a viscous liquid. Phys Fluids 1:338–346
17. Brodnyan JG (1959) The concentration dependence of the Newtonian viscosity of prolate ellipsoids. Trans Soc Rheol 3:61–68

18. Boussinesq J (1913) Vitesse de la chute lente, devenue uniforme, d'une goutte liquide spherique, dans un fluide visqueux de poids specifique moindre. Comp Rend Acad Sci 156:1124–1129
19. Bowen RM (1967) Towards a thermodynamics and mechanics of mixtures. Arch Rational Mech Anal 24(5):370–403
20. Bowen RM, Garcia DJ (1970) On the thermodynamics of mixtures with several temperatures. Int J Eng Sci 8:63–83
21. Bowen RM (1991) Theory of mixtures. In: Continuum physics, Ed Eringer AC, Academic Press, New York, Vol III
22. Burgers JM, Jeffery GB (1939) Mechanical considerations – Model systems – Phenomenological theories of relaxation and of viscosity, First Report on Viscosity and Plasticity, 2nd edn. Nordemann Publishing Company, Inc., New York Prepared by the committee for the study of viscosity of the academy of sciences at Amsterdam. Proc R Soc Lond A 102:161–179
23. Craig RE (1970) The motion of a solid in a fluid. Am J Math 20:162–177
24. Craine RE, Green AE, Naghdi PM (1970) A mixture of viscous elastic materials with different constituent temperatures. Quart J Mech Appl Math 23:171–184
25. Darcy H (1856) Les Fontaies Publiques de La Ville de Dijon. Victor Delmont
26. Einstein A (1906) Eine neune Bestimmung der Molekul-Dimension. Ann Phys 19:289–306
27. Einstein A (1911) Eine neune Bestimmung der Molekul-Dimension. Ann Phys 34:591–592
28. Fick A (1855) Uber diffusion. Ann Phys 94:59–86
29. Green AE, Naghdi PM (1969) On basic equations for mixtures. Quart J Mech Appl Math 22:427–438
30. Greenhill AG (1898) The motion of a solid in infinite liquid under no forces. Am J Math 20:1–75
31. Hadmard JS (1911) Mecanique-mouvement permanent lent une sphere liquids et visqueuse dans un liquide visqueux. Comp Rend Acad Sci 154:1735–1738
32. Hadmard JS (1912) Hydrodynamique – Sur une question relative aus liquids visqueux. Comp Rend Acad Sci 154:109
33. Happel M, Brenner H (1973) Low Reynolds number hydrodynamics, Noordhaff, Leyden
34. Hatschek E (1928) The viscosity of liquids. Bel G and Son, London
35. Johnson G, Massoudi M, Rajagopal KR (1991) Flow of a fluid infused with solid particles through a pipe. Int J Eng Sci 29(6):649–661
36. Jeffery GB (1922) The Motion of ellipsoidal particles immersed in a viscous fluid. Proc R Soc Lond A, Containing papers of a mathematical and physical character 102(715):161–179 http://www.jstor.org/stable/94111
37. Kirchoff G (1869) Uber die benegung eines rotationskorpers in earner fliissigkeit. Crelle 71:237–262
38. Lamb H (1877) On the free motion of a solid through an infinite mass liquid. Math Soc 8: 273–286
39. Malek J, Rajagopal KR (2008) A thermodynamic framework for a mixture of two liquids. Nonlinear Anal R World Appl 9(4):1649–1660
40. Massoudi M (1986) Application of mixture theory to fluidized beds, PhD thesis. University of Pittsburgh
41. Mills N (1966) Incompressible mixtures of Newtonian fluids. Int J Eng Sci 4:97–112
42. Mooney MJ (1954) On an indeterminate integral in Einstein's theory of the viscosity of a suspension. J Appl Phys 25:406–407
43. Munaf DR, Lee D, Wineman AS et al (1993) A boundary value problem in groundwater motion analysis – Comparison of predictions based on Darcy's law and the continuum theory of mixtures. Math Models Methods Appl Sci 3:231
44. Oldroyd JC (1953) The elastic and viscous properties of emulsions and suspensions. Proc R Soc Lond A 218:122–132
45. Rajagopal KR, Tao L (1992) Wave propagation in elastic solids infused with fluids. Int J Eng Sci 30:1209–1232
46. Rajagopal KR, Tao L (1995) Mechanics of mixtures, World Scientific, Singapore

47. Rajagopal KR (2003) Diffusion through solids undergoing large deformations. Mater Sci Technol 19:1175–1189
48. Rajagopal KR, Srinivasa AR (2004) On thermomechanical restrictions of continua. Proc R Soc Lond Ser A: Math Phys Eng Sci 460(2042):631-651
49. Rajagopal KR, Srinivasa AR (2004) On the thermomechanics of materials that have multiple natural configurations – Part I: Viscoelasticity and classical plasticity. Zeitschrift für Angewandte Mathematik und Physik 55(5):861–893
50. Rajagopal KR, Srinivasa AR (2004) On the thermomechanics of materials that have multiple natural configurations – Part II: Twinning and solid to solid phase transformation. Zeitschrift für Angewandte Mathematik und Physik 55(6):1074–1093
51. Rajagopal KR, Srinivasa AR, A Gibbs-potential-based formulation for obtaining the response functions for a class of viscoelastic materials, Proceedings of the Royal Society – A Mathematical, Physical and Engineering Sciences, Published online before print June 16, 2010, doi:10.1098/rspa.2010.0136
52. Rayleigh JWS (1892) Correlation aspects of the viscosity–temperature relationship of the lubricating oils, PhD thesis, Technische Hogeschool Delft, The Netherlands
53. Rybczynski W (1911) On the translatory motion of a fluid sphere in a viscous medium. Bull Acad Sci Cracow Ser A:447–459
54. Saltzer WD, Schulz B (1966) An attempt to treat the viscosity as a transport property of two phase materials. In: Continuum models of discrete systems 4, Eds. Erwin O and Hsieh RKT, North-Holland, Amsterdam
55. Samohyl I (1987) Thermodynamics of irreversible processes in fluids mixtures, Teubner, Leipzig
56. Seitz F (1987) The modern theory of solids. Dover Publications, New York
57. Simha R (1952) A treatment of the viscosity of concentrated suspensions. J Appl Phys 23:1020–1024
58. Stokes GG (1845) On the effect of the internal friction of fluids on the motions of pendulums. Trans Camb Phil Soc 8:287–305
59. Tamura M, Kurata M (1952) On the viscosity of binary mixture of liquids. Bull Chem Soc Jpn 25(1):32–38
60. Taylor GI (1932) The viscosity of fluids containing small drops of another fluid. Proc R Soc Lond A 138:41–48
61. The Oxford English Dictionary (1981) Oxford University Press, Oxford
62. Thomson W, Tait PG (1867) Treatise on natural philosophy, Cambridge University Press
63. Truesdell C (1957) Sulle basi della thermomeccanica. Rend Lincei 22:33–38
64. Truesdell C (1957) Sulle basi della thermomeccanica. Rend Lincei 22:158–166
65. Truesdell C (1991) A first course in rational continuum mechanics, Academic Press, New York
66. Truesdell C (1984) Rational thermodynamics, Springer-Verlag, Berlin
67. Quemada O (1977) Rheology of concentrated disperse systems and minimum energy dissipation principle, I. Viscosity–concentration relationship. Rheol Acta 16:82–94
68. Williams WO, Sampaio R (1977) Viscosities of liquid mixtures. Z Angew Math Phys 28:607-614
69. Ziegler H (1963) Some extremum principles in irreversible thermodynamics. In: Sneddon IN, Hill R, Eds, Progress in solid mechanics vol. 4, North Holland Publishing Company, New York
70. Ziegler H (1983) An introducton to thermomechanics, North Holland Publishing Company, Amsterdam, 2nd Edn
71. Ziegler H, Wehrli C (1987) The derivation of constitutive equations from the free energy and the dissipation function. In: Wu TY, Hutchinson JW, Eds, Adv Appl Mech, Academic Press, New York 25:183–238

Part II
Rheology

Chapter 4
Oscillatory Shear Rheology for Probing Nonlinear Viscoelasticity of Complex Fluids: Large Amplitude Oscillatory Shear

Abhijit P. Deshpande

Abstract Linear and non-linear rheological responses of complex fluids are of great interest. Oscillatory techniques are commonly used to analyze the complex fluid rheological behavior. In this chapter, the approach based on large amplitude oscillatory shear (LAOS) is reviewed. Initially, oscillatory shear based on small strain amplitudes is presented along with a brief discussion on relaxation time spectrum. Subsequently, key observable features of LAOS are shown with experimental observations on selected materials. Various applications for which LAOS is being investigated have been described through these examples.

Complex fluids such as polymer solutions and melts, emulsions, blends, biological gels, micellar solutions, etc. are being investigated for their non-linear response at large amplitudes during LAOS. In addition to the general response during LAOS, specific material functions being proposed in the literature are discussed. Finally, an example of bulk oscillatory flow is discussed in the context of LAOS behavior of complex fluids.

4.1 Introduction

Rheological characterization is carried out using several *simple* controlled methods, such as steady shear, stress relaxation, creep, oscillatory shear and steady extension. The results of these tests are quantified using material functions such as steady viscosity, relaxation modulus, creep compliance, storage and loss modulus and extensional viscosity, respectively. For specific materials, other rheological tests have been used frequently. These tests such as double step shear, stress growth and superimposed oscillations on prescribed strain or stress are more appropriate considering the engineering application or they serve to highlight the effect of molecular or microstructural features on material response more effectively.

A.P. Deshpande (✉)
Indian Institute of Technology Madras, Chennai 600036, India
e-mail: abhijit@iitm.ac.in

Oscillatory shear is used widely in characterization of viscoelastic materials [2, 16]. In this method, both stress and strain vary cyclically with time, with sinusoidal variation being the most commonly used. This is the most popular method to characterize viscoelasticity, since relative contributions of viscous and elastic responses of materials can be measured. The cycle time, or frequency of oscillation, defines the timescale of these tests. By observing material response as a function of frequency, the material can be probed at different timescales. This observation of material response at different frequencies is also referred to as mechanical spectroscopy. The overall material response is due to contributions from several mechanisms at the molecular and microscopic levels. A set of timescales can be identified with each mechanism. The ratio of the two timescales, the experimental and material, can be varied by observing response at different frequencies.

4.1.1 Small Amplitude Oscillatory Shear

We can represent the sinusoidal strain applied to a complex fluid sample as

$$\gamma = \gamma_0 \sin\omega t. \tag{4.1}$$

The *linear* response of material in terms of stress can be written as

$$\sigma = \sigma_0 \sin(\omega t + \delta), \tag{4.2}$$

where δ is the phase lag. The response given by the above equation is usually observed at low amplitudes of strain (γ_0). At larger strain and/or stress amplitudes, the *nonlinear* response of materials is discussed later. Different waveforms, such as triangular, square and trapezoidal have also been used in oscillatory shear.

In small amplitude oscillatory shear (SAOS), based on the strain imposed and the stress response, material functions are defined to quantify the material behavior. For example, *storage* (G') and *loss* (G'') moduli are defined as ratios of stress and strain amplitudes. Storage modulus is based on the amplitude of in-phase stress and loss modulus is based on the out-of-phase stress. Based on these material functions, (4.2) can be written as

$$\sigma = G'(\omega)\sin\omega t + G''(\omega)\cos\omega t. \tag{4.3}$$

A primitive model to describe viscoelastic behavior of materials is the Maxwell model,

$$\sigma + \lambda \frac{\partial \sigma}{\partial t} = \eta \dot{\gamma}, \tag{4.4}$$

where λ (the relaxation time) and η are the model parameters. λ is the characteristic time of material response. Large values of λ imply elastic response, while small values imply viscous response. η is the characteristic zero shear viscosity. When subjected to SAOS, the response of Maxwell model is

$$G' = \frac{\eta \omega^2 \lambda}{1 + \omega^2 \lambda^2}, \quad G'' = \frac{\eta \omega}{1 + \omega^2 \lambda^2}. \tag{4.5}$$

As can be observed from the above equation, material response at very low frequencies is $G' \propto \omega^2$ and $G'' \propto \omega$, signifying viscous response. At very high frequencies, material response is $G' = \eta/\lambda$ and $G'' \propto \omega^{-1}$. The constant value of storage modulus is G ($= \eta/\lambda$), and is called the elastic modulus. The crossover between G' and G'' occurs at the frequency $\omega = 1/\lambda$. A complex modulus, $\mid G^* \mid = \sqrt{G'^2 + G''^2}$, is also sometimes used for analyzing the behavior. Using complex notation, applied sinusoidal strain can be written as

$$\gamma = \gamma_0 e^{i\omega t}. \tag{4.6}$$

If the above strain is applied and for the case of linear response, one can write

$$\sigma = G^* \gamma. \tag{4.7}$$

As we will see later, the above equation can be seen as the truncated form (upto the linear term) of the overall relation between stress and strain.

Either stress or strain can be visualized as the *excitation* and consequently, either strain or stress can be visualized as the *response*. The results of oscillatory shear tests are presented in different ways for analyzing the data. An example is plot of G'' vs. G' or the Cole–Cole plot. Response of Maxwell model, as given by (4.5), on Cole–Cole plot is represented by a semi-circle with centre at (G/2,0) and radius of G/2. Plots of G^* vs. δ are also used to analyse the property variation.

The cyclic variation of stress and strain can be observed by plotting them against each other. For fluids, stress and strain rate are also plotted. Examples of stress–strain variation, also called *Lissajous plots*, are shown in Fig. 4.1. Stress and strain are in phase for elastic materials (maximum stress when strain is maximum, etc.)

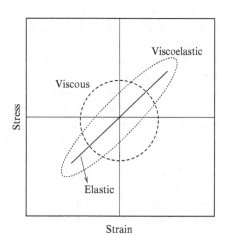

Fig. 4.1 Stress–strain during a cycle for different materials

leading to a straight line in case of linear elastic materials. On the other hand, they are out of phase ($\pi/2$) for viscous materials (implying maximum stress when strain rate is minimum). This behavior is represented by a circle in the plot, if the fluid is linear viscous or Newtonian. It should be noted that this response would be a circle or an ellipse (with major and minor axes coinciding with stress and strain axes), depending on the scales used. For a viscoelastic material, with a phase lag $0 < \delta < \pi/2$, the stress–strain curve is elliptical with major and minor axes not coinciding with stress and strain axes.

Similar plots of stress–strain rate would result in a straight line for linear viscous response, circle for linear elastic response and an ellipse for linear viscoelastic materials, respectively.

The linear elastic and linear viscous responses are represented by straight lines or circles. For non-linear elastic or non-linear viscous materials, these plots would be non-linear curves instead of straight lines or they would be ellipses (with major and minor diameters coinciding with axes) instead of circles. Before discussing such response for non-linear viscoelastic behavior, typical examples of response to SAOS are discussed.

4.1.2 Typical Response

Example variations of G' and G'' are shown in Fig. 4.2 for a polymer melt, an emulsion and a crosslinking polymer. For the polymer melt (Fig. 4.2a, also shown in Fig. 8.2), at very low frequency, viscous behavior is observed. At higher frequencies, the behavior is largely governed by the entanglements between polymer molecules. This region is also referred to as the plateau region, due to relatively constant moduli. At very high frequencies, the response is almost elastic. This response is also referred to as the glassy behavior. It should be noted that the change from viscous to elastic behavior is observed over a couple of decades in frequency (a change of 100) in case of Maxwell model. However, for most materials this change, if at all observed, occurs over several decades. For polymer melts, the moduli change by 4 orders of magnitude for a 7–8 orders of magnitude change in frequency.

The rheology of the emulsions is strongly influenced by the state of flocculation [9]. Unflocculated or weakly flocculated emulsions show a crossover point between G' and G''. This crossover frequency is associated with a characteristic relaxation time for the onset of the terminal or flow region for the emulsion. Highly concentrated stabilized emulsions show a gel-like response, implying G' to be larger than G'' and both being almost constant with respect to frequency. Figure 4.2b shows the normalized response of emulsions (with different stabilizer concentrations). As mentioned, G' is larger than G''. Both change by an order of magnitude in the range of frequencies investigated [9]. Therefore, a plateau-like region can be observed as the overall response. Similar to the entanglement plateau region in case of polymer, this region in emulsions may be due to interactions between emulsifiers from neighbouring droplets [9].

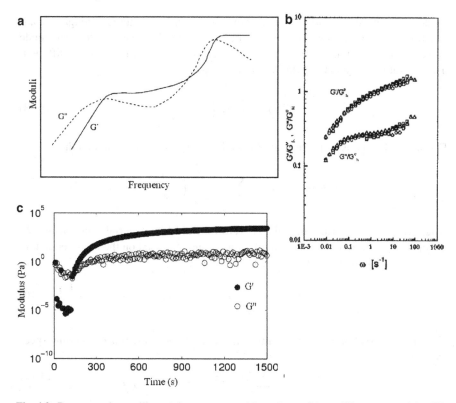

Fig. 4.2 Representative oscillatory shear response: (**a**) a polymer (**b**) an oil in water emulsion [9] (**c**) evolution during gelation [10]

Figure 4.2c shows evolution of linear viscoelastic response for a sample under-going gelation. In this case, sodium acrylate is being crosslinked in the presence of free-radical crosslinker. Initially, G' is observed to be lower than G''. With crosslinking reaction and network formation, both G' and G'' increase. Near the gel point, there is a crossover in G' and G'' and subsequently G' is larger than G'' [10]. Once the reaction is complete and gel is formed, both G' and G'' become constant. G' and G'' for the gel are almost independent of frequency.

The overall response to SAOS, as exemplified in Fig. 4.2, is very complex compared to the simplistic response exhibited by the Maxwell model, which is characterized by a single relaxation time. We understand the overall response by analysing it to be due to combinations of several relaxation times.

4.1.3 Relaxation Time Spectrum

The overall material response is visualized in terms of combinations of several mechanisms and modes. Each of the modes is described in terms of a Maxwell model. Therefore, overall response of material can be captured through a combination of Maxwell models (or generalized Maxwell model) with each mode corresponding to a relaxation time. The strength of each mode may also be different. The relaxation modulus for Maxwell model is given by

$$G(t) = G \exp\left(-\frac{t}{\lambda}\right). \tag{4.8}$$

For material response with several Maxwell models, the relaxation modulus is

$$G(t) = G_e + \sum_i G_i \exp\left(-\frac{t}{\lambda_i}\right), \tag{4.9}$$

where G_e, the equilibrium modulus, is zero for fluid-like materials, and G_i and λ_i are elastic modulus and relaxation time for ith mode. Based on the above equation, we can define a relaxation time spectrum, $H(\lambda)$, as

$$H(\lambda) = \sum_i G_i \delta(\lambda - \lambda_i). \tag{4.10}$$

Therefore, relaxation modulus can be written in terms of the relaxation time spectrum as

$$G(t) = G_e + \sum_i G_i \delta(\lambda - \lambda_i) \exp\left(-\frac{t}{\lambda}\right). \tag{4.11}$$

Similarly, we can define the relaxation modulus based on the continuous relaxation time spectrum as

$$G(t) = G_e + \int \frac{H(\lambda)}{\lambda} \exp\left(-\frac{t}{\lambda}\right) d\lambda. \tag{4.12}$$

The storage and loss moduli can be written in terms of relaxation time spectrum as follows:

$$G' = \sum_i \frac{G_i \omega^2 \lambda_i^2}{1 + \omega^2 \lambda_i^2}, \quad G'' = \sum_i \frac{G_i \omega \lambda_i}{1 + \omega^2 \lambda_i^2}. \tag{4.13}$$

Alternately, in terms of continuous spectrum,

$$G' = \int \frac{H(\lambda)\omega^2 \lambda}{1 + \omega^2 \lambda^2} d\lambda, \quad G'' = \int \frac{H(\lambda)\omega}{1 + \omega^2 \lambda^2} d\lambda. \tag{4.14}$$

Based on these equations, relaxation time spectrum can be estimated from data, such as in Fig. 4.2, and the mechanisms of material behavior can also be understood in terms of the distribution of relaxation times.

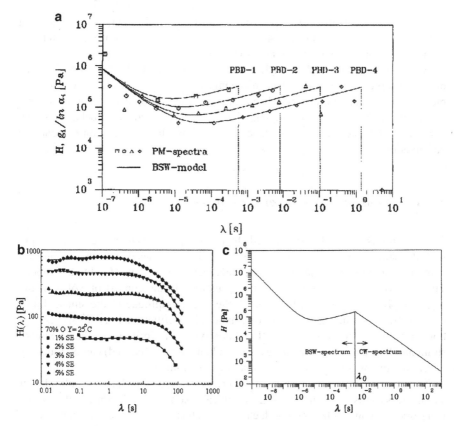

Fig. 4.3 Relaxation time spectrum of (**a**) monodisperse polybutadienes [1] (**b**) emulsion [9] (**c**) a gel near gelation threshold [27]

An example of discrete relaxation time spectrum (referred to as Parsimonious model (PM) spectrum in the figure) evaluated from oscillatory shear data is shown in Fig. 4.3. These data are for different molecular weights of monodisperse polybutadiene samples. Analytical expressions for relaxation time spectrum have also been proposed and one example is (Baumgärtel, Schausberger and Winter (BSW) spectrum in Fig. 4.3) [1]

$$H(\lambda) = H_e \lambda^{n_e} + H_g \lambda^{-n_g} \qquad \lambda_1 < \lambda < \lambda_{\max}$$
$$= 0 \qquad\qquad\qquad\qquad \lambda > \lambda_{\max}, \qquad\qquad (4.15)$$

where λ_1 is the shortest measurable relaxation time and $\lambda_{\max}$ is the largest relaxation time of the polymer. When tested below $1/\lambda_{\max}$ (i.e., at strain rate or frequency lower than this value), the polymer would show Newtonian viscous behavior. H_e, n_e and H_g, n_g are coefficients to capture the entanglement modes and glassy modes of the polymer, respectively.

Examples of relaxation time spectrum for polymers and an emulsion are given in Fig. 4.3. As was observed in Fig. 4.2 with G' and G'', there is a marked difference in the relaxation time spectra for both the materials.

As discussed in Sects. 4.1.1–4.1.3, material response can be described in terms of microscopic mechanisms. These mechanisms for complex fluids, such as polymers, colloids, gels, liquid crystals and micelles, are recognized based on microscopic/molecular cooperative *organization* [17] . The viscoelastic behavior in the linear and non-linear regimes is dependent on how the organization responds to deformation.

4.1.4 Linear and Nonlinear Responses

The response is termed linear if scaled change in input leads to change in the output with the same scaling. Describing linear response in other words, the output of a combination of inputs is the same as the combination of outputs of individual inputs. This is also referred to as superposition principle. In the context of SAOS, if the strain amplitude is changed by a factor, the stress amplitude of the sinusoid also changes by the same factor. Therefore, material functions such as G' and G'' are not functions of strain amplitude. The linear viscoelastic limit (or the maximum strain amplitude at which linear viscoelastic behavior is observed) can be found by measuring the material functions as functions of strain amplitude.

The onset of nonlinear response is therefore expected at larger amplitudes of strain/stress as discussed in Sect. 4.2. Nonlinear response, in the context of rheological response, can be classified as due to large deformations, structural changes and phase transitions [18]. Analysis of nonlinear response is complicated because interplay of these factors is difficult to resolve, and their mathematical representations are difficult to propose and solve. However, in the recent decades, a lot of progress in theoretical and experimental tools has led to significant understanding. The use of oscillatory shear in the nonlinear regime is an example of such endeavor.

As in the case of linear viscoelastic behavior, several methods can be used to examine the nonlinear response of the materials. These include creep and recovery, stress relaxation and oscillatory shear. These different methods serve to highlight particular structure–property relations or they are more relevant for an application. Along with experimental characterization, development and use of comprehensive models that explain behavior of different materials in various methods of probing are continuously being undertaken.

4.1.5 Susceptibility

Linear and nonlinear responses of materials to excitation are of interest in various fields, such as mechanical, electrical, thermal and their combinations. The material property relating the response and excitation is referred to as susceptibility. At low levels of excitation, linear response is observed. In other words, the implication is (in case of oscillatory excitation) that the susceptibility is independent of the amplitude of excitation.

Linear response is usually very important in understanding basic mechanisms responsible for material behavior. Nonlinear response, on the other hand, is more relevant for applications and is also more difficult to characterize. Measuring the nonlinear response of a material to an excitation is a way to examine properties that cannot be characterized by examining the linear response. Understanding nonlinear effects has led to breakthroughs in different materials, such as elastic, plastic, viscoelastic, optical materials, ferroelectric, freezing, or dipolar glass transitions, isotropic-liquid crystal transition or binary mixtures, superconductivity, field or heating effects in electrical transport and heating due to electric field excitation of supercooled liquids [24].

Similar techniques are used for the analysis of nonlinear response in these diverse areas. As an example, the response in dielectric spectroscopy and mechanical spectroscopy can be understood in relation to each other. For small amplitude of electric field, similar to (4.7), we can write the relation between electric displacement (D) and electric field (E) as

$$D = \epsilon^* E, \tag{4.16}$$

where ϵ^* is permittivity of material. The modes of material response, as discussed in earlier sections, can also be identified by examining the dielectric response as a function of frequency. In case of rheology, load and displacement are measured and analysis is carried out for stress, strain or strain rate. In case of dielectric response, we measure current and voltage and carry out the analysis with electric field, polarisation or electric displacement.

When the material is subjected to a large amplitude of electric field, the above equation is no longer valid. However, material behavior can be described by writing higher order terms in electric field [3],

$$D = \epsilon_1^* E + \epsilon_2^* E^2 + \epsilon_3^* E^3 + ..., \tag{4.17}$$

where ϵ_n is the permittivity of nth order. It can be shown that only the odd powered terms of the above equation are non-zero. In addition to such general descriptions, a variety of theoretical and experimental tools is common in investigations of nonlinear response of materials. In the next few sections, various features of nonlinear response in oscillatory shear will be described.

4.2 Oscillatory Testing at Large Amplitude

The classes of overall oscillatory rheological response of materials can be understood from the diagram shown in Fig. 4.4. The diagram (referred to as *Pipkin diagram*) can be recast in the form of dimensionless numbers, Weissenberg number ($We = \dot{\gamma}\lambda$) and Deborah number ($De = \omega\lambda$). At low frequencies, and at low

Fig. 4.4 Diagram showing
material response at different
frequencies and strain rates

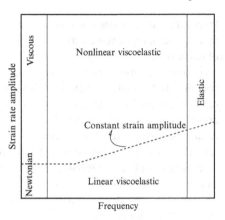

strain amplitudes, the material response is purely viscous and Newtonian. As
mentioned earlier, this is at timescales larger than the largest relaxation time in
the material. With an increase in frequency, we observe viscoelastic response at
timescales smaller than the largest relaxation time in the material. Depending on
the strain amplitude or strain rate amplitude, a crossover from linear viscoelastic to
nonlinear viscoelastic is observed. At low frequencies, this crossover occurs beyond
a threshold of strain rate amplitude, while at high frequencies, it occurs beyond
a threshold of strain amplitude. At very high frequencies, with timescales shorter
than the smallest relaxation time of the material, all modes are frozen and an elastic
response is observed.

The most common nonlinear viscoelastic measurement is steady shear viscosity
(where at different strain rates, steady stress response of the material is measured).
Using this, we can distinguish different types of behavior, such as shear thinning
and shear thickening. The material behavior, as captured by steady viscosity, is not
dependent on the direction of rotation; $\eta = \eta\left(\dot{\gamma}\right) = \eta\left(-\dot{\gamma}\right)$.

The most common linear viscoelastic measurement is based on oscillatory shear,
as defined in Sect. 4.1. Most engineering applications may involve neither steady
flow nor small deformation. Therefore, large amplitude oscillatory shear (LAOS)
is suggested to investigate transient behavior of the material at large deformations.
It can also be used to quantify the progressive transition from linear to nonlinear rheological behavior. At larger amplitudes of strain, oscillatory shear response will be
nonlinear viscoelastic and has been investigated for the last couple of decades. However, only in the recent few years, LAOS is being used to elucidate specific features
of different materials. Some of these examples are described briefly in Sect. 4.4.

4.2.1 Qualitative Description During LAOS

In SAOS, the material response, in terms of stress, is periodic as given in (4.2). The
strain amplitude limit for observation of linear viscoelastic response (below which
SAOS is usually performed) is small. For many materials, it is less than or around 1.

When the material is subjected to LAOS, the overall stress response has been observed to be periodic as well. In both SAOS and LAOS, a certain number of cycles is required before the *terminal steady* behavior is observed (implying cyclic response to be same for nth and $(n + 1)$th cycles). The analysis, in both cases, is carried out once the terminal steady behavior is reached.

The overall frequency of the terminal periodic response, in case of LAOS as well, is largely the input frequency of the strain. When we take a closer look at the cycle, the stress sinusoid may be *narrower* or *wider* near the peaks when compared to response if a single frequency response were observed. In addition, the symmetries before and after the peak may not be present [8]. Similar features are shown when the material is subjected to a given stress and strain response is observed [20]. This qualitative response implies that stress response is not governed only by the sinusoidal behavior at the input strain frequency. Therefore, stress response can be visualized as being composed of various frequencies. The breaking of the before and after stress symmetries also implies that phase differences exist among responses at various frequencies.

An example of sinusoidal response during LAOS is shown in Fig. 4.5a, b. Although all the features mentioned above are shown in the figure, the response, at first glance, may not seem very different. Indeed, the effect of various experimental errors has to be carefully considered while analyzing the LAOS data. This is discussed further in Sect. 4.2.4.

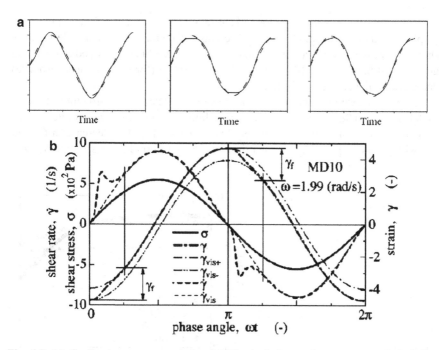

Fig. 4.5 (a) Cyclic stress response during LAOS, single harmonic response as *dashed line* (b) example strain response under LAOS for concentrated suspension [20]

Figure 4.5 also shows another example of deviation from SAOS cyclic response. In this type of response (measured for a concentrated suspension [20]), there is a departure from SAOS behavior during part of the cycle. Significant deviations in strain rate are observed during reversal of flow direction, i.e., when strain starts increasing from the minimum value or when strain starts decreasing from maximum value.

Based on this basic description of cyclic response during an LAOS experiment, the following features can be used to quantify the material response:

Amplitudes of Stress and Strain The magnitude of first harmonic as a function of strain amplitude or ratio of stress and strain amplitudes can be examined for storage and loss moduli. These properties capture the material response at the same frequency as the input frequency. These measurements are routinely made to find the range of linear viscoelastic response of a material. In these measurements, one can visualize stress response to be written in the form of

$$\sigma = \sigma_0 \left(\gamma_0 \right) \sin \left(\omega t + \delta \right). \tag{4.18}$$

The above equation states that material response to LAOS is modified (when compared to that to SAOS) by only the amplitude of stress being dependent on the strain amplitude. The sinusoid is represented by the same function in SAOS and LAOS. When $\sigma_0 \sim \gamma_0$ constant, linear viscoelastic response or SAOS behavior is observed.

The onset of nonlinear behavior and subsequent variation as a function of strain amplitude has been examined for a large class of materials. An example result is shown for poly sodium acrylate gel and cellulose microfibrils/poly sodium acrylate gel in Fig. 4.6. It can be observed that the onset strain (for the nonlinear response) decreased in the presence of microfibrils.

As another example, the onset strain has been related to the structure of a colloidal gel [23]. The structure of polymeric solutions and intermolecular interactions in them can be distinguished based on this response [13]. The decrease in G' and G'' with strain amplitude, as shown in Fig. 4.6, is referred to as *strain softening*.

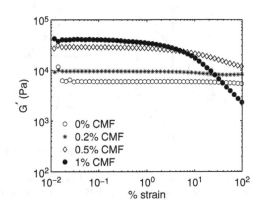

Fig. 4.6 G' and G'' as functions of strain amplitude for gels [10]

In addition, depending on the material, one can observe increase in G' and G'' (*strain hardening*) and overshoots in G' and G'' followed by decrease [13]. These changes correspond to changes in stress amplitude $\sigma_0(\gamma_0)$ (4.18).

Stress vs. Strain During a Cycle: Lissajous Plot As shown in Fig. 4.1, the stress–strain curves for viscoelastic material subjected to SAOS would be ellipses with different major and minor axes, depending on the relative contributions of viscous and elastic responses (in other words, depending on amplitude of stress response and its phase difference with strain).

The stress–strain curves in case of LAOS are departures from ellipses. This can be understood based on the qualitative features shown in Fig. 4.1. The departure from linear response (or the response at the input frequency) is exhibited as stretched and deformed ellipses. The response being considered is stress–strain, and therefore this departure is described again in terms of *softening* and/or *hardening* of the material during a cycle.

LAOS response can therefore be examined through stress–strain or stress–strain rate plots. Examples of these are shown in Fig. 4.7. The plot shown in Fig. 4.7a is

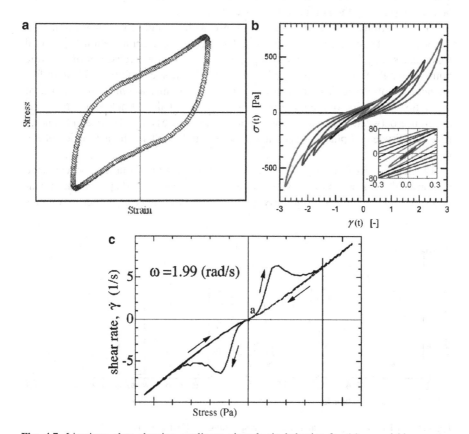

Fig. 4.7 Lissajous plots showing nonlinear viscoelastic behavior for (**a**) a vanishing cream (**b**) mucus gel [7] (**c**) departure from linear viscous behavior for a concentrated suspension [20]

for a vanishing cream. As can be seen, there is a significant departure from linear viscoelastic response (or elliptical plot). At larger strains in a cycle, proportionally larger stress is required for higher strain and therefore this behavior is referred to as strain hardening during the cycle.

Stress–strain curves for mucus gel [7] at different strain amplitudes are shown in Fig. 4.7b. Variation of G' and G'', for the various strain amplitudes, was observed to be minimal. Therefore, little departure from linear response as described in the first feature was observed. However, when Lissajous plots are observed (Fig. 4.7b), one can notice a striking departure from linear viscoelastic behavior. Only at low amplitudes (shown in inset) can one observe the elliptical plots.

Another example of cyclic response during LAOS is shown for a concentrated particulate suspension [20]. The data, which correspond to Fig. 4.7c, have been shown with stress–strain rate plot. A linear plot would indicate viscous response, with stress and strain rate in phase. As mentioned earlier, departure from this behavior is observed during certain duration of the cycle.

Frequency Spectrum Fourier transform of the time domain signal is used to evaluate the frequency spectrum. The magnitudes of the peaks at higher harmonics can be used to quantify the nonlinear response of materials.

The deviations from a sinusoidal response with a single frequency (input frequency) can be captured by analysing the signal as a combination of several frequencies [25]. This is shown in Fig. 4.8 with the Fourier transform of the time domain signal. As will be discussed later, higher harmonics (multiples of input frequency) are observed. The intensities of higher harmonics are much less than that of the first harmonic (or the input frequency). It should also be noted that the phase difference between strain and various harmonics of stress can be different. Therefore, amplitude and phase of each harmonic are independent characteristics of the material.

The dependence of G' and G'' on strain amplitude signifies nonlinear viscoelastic response. The distortion of stress–strain loops (from elliptical shapes) is an indication of nonlinear behavior. Similarly, the presence or absence of higher harmonics

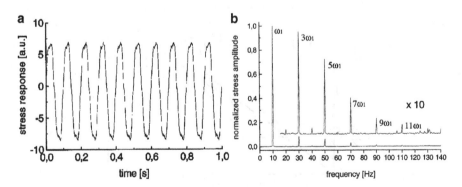

Fig. 4.8 (a) Time domain signal and the corresponding (b) Fourier transform [25]

in the material response can be used as an indicator of nonlinear or linear response. However, it is also important to get quantitative measures from all of these measurements. A couple of the quantitative measures being proposed as *nonlinear material functions* are discussed in Sect. 4.2.3. In addition, it is important to relate the nonlinear response as captured by LAOS to physicochemical processes in the material. Therefore, simulation of LAOS response with different constitutive models is also an important area of activity.

4.2.2 General Response in LAOS

When a material is subjected to strain as given in (4.1), the general stress response of the material can be written as

$$\sigma(t) = \sum_{1}^{N} \sigma_n \sin(n\omega t + \delta_n), \tag{4.19}$$

where σ_n and δ_n are amplitude and phase lag of the nth harmonic. In this equation, σ, a function of time, is being expressed as a Fourier series. In case of linear viscoelasticity or SAOS, only the first term of this series exists (4.2).

The expression for stress can also be written as a series using complex notation,

$$\sigma = \sum_{1}^{N} G_n^* \gamma^n. \tag{4.20}$$

For linear viscoelastic response, only the first term of the summation exists (4.7). Similar to storage and loss moduli in case of linear viscoelastic response, we can define a series of storage and loss moduli G_n' and G_n'', corresponding to nth harmonic in the series in (4.20). These moduli can also be written in terms of σ_n and phase δ_n from (4.19).

For isotropic liquids and given that viscosity is not a function of direction of strain rate, it can be shown that only the odd harmonics of the above series are non-zero. This is also the case for dielectric properties, where dielectric constants are not functions of the direction of electric field. The overall number of terms, N, can be arbitrarily large. However, most measurements have been made upto the fifth harmonic due to experimental limitations. The issues related to detection of higher harmonics are discussed later in Sect. 4.2.4. Fourier transform of $\sin \omega_0 t$ is delta functions at ω_0. Therefore, Fourier transform of (4.19) will be a set of peaks corresponding to each harmonic (as shown in Fig. 4.8b).

The measurement of stress, through the torque or force sensors of the rheometer, is done discretely at a finite sample interval (t_s). Based on the sampling frequency ($N_s = 1/t_s$, number of data points per second), we obtain a time series of discrete measurements. This time domain series can be denoted as

$$\sigma(t) = \sum_{1}^{N_s T} \sigma(k)\delta(t - t_s k), \tag{4.21}$$

where k takes integer values from 1 to $N_s T$ and T is the total duration of measurement. Alternately, the series can be written as $\sigma(k t_s)$, where $k = 1, 2, 3...$

Discrete Fourier transform of this time domain series will also be a series of $N_s T$ complex numbers. The maximum frequency in the Fourier domain corresponds to Nyquist frequency $= 2\pi(1/2t_s)$. With the property of Fourier transform leading to meaningful N/2 terms (symmetric), the resolution in frequency domain is $2\pi(1/2t_s) \times (1/(N_s T/2)) = 2\pi(1/T)$. Therefore, sampling interval determines maximum frequency to which information can be obtained, while duration of measurement determines the resolution of frequency. Larger T also leads to higher signal to noise ratio [26].

In the Fourier domain, the intensity of nth harmonic peak is indicated by I_n. The algorithm for identification of higher harmonics has to consider the resolution in the frequency domain, as the frequency spectrum will not have frequencies, which are integral multiples of each other.

This general description will be useful for understanding material behavior if we can propose quantitative measures based on the evaluated variables. Although LAOS is being investigated for a couple of decades, only in recent times do we have the theoretical and experimental tools to attempt quantifying the response. In the next section, a couple of examples of such quantification are discussed.

4.2.3 Proposed Material Functions in LAOS

The moduli G'_n as defined in previous section are based on the nth harmonic. Whenever we analyze one of these moduli, we do not examine the overall response, but look at overall response as contributions from these individual moduli. Based on the time series of stress, moduli can be defined which are not based on an individual harmonic alone [7]:

$$G'_M = \left(\frac{d\sigma}{d\gamma}\right)_{\gamma=0},$$

$$G'_L = \left(\frac{\sigma}{\gamma}\right)_{\gamma=\gamma_0}. \tag{4.22}$$

These moduli will reduce to G' in case of SAOS (linear viscoelasticity). Additionally, they have geometric interpretation as shown in Fig. 4.9. In nonlinear elasticity, tangent modulus and secant modulus are used very commonly to describe the stiffness of a material. Analogous to these, G'_M is the tangent modulus at zero strain and G'_L is the secant modulus at maximum strain. In case of linear viscoelastic response, using (4.1) and (4.2), we can show that $G'_M = G'_L = G'$ and

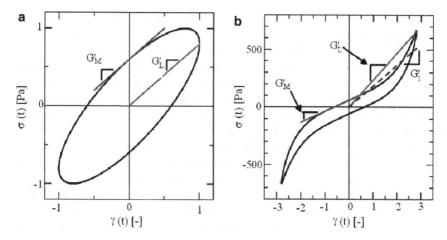

Fig. 4.9 Proposed elastic moduli and their geometric interpretation for (**a**) linear and (**b**) nonlinear viscoelastic behavior [7]

this is shown geometrically in Fig. 4.9a. For nonlinear viscoelastic response, these moduli capture the relative strain softening/hardening that occurs during one cycle and this is depicted in Fig. 4.9b. As the figure shows, due to strain hardening, the secant modulus G'_L is larger than the tangent modulus G'_M.

Other approaches are based on the ratios of intensities of higher harmonics and first harmonic of the stress response. For example, coefficient Q is defined as [12]

$$Q = \frac{1}{\gamma_0^2} \frac{I_3}{I_1}. \tag{4.23}$$

This coefficient has been claimed to be helpful in distinguishing molecular architecture of polymers based on LAOS.

Through analysis of shapes in Lissajous plots and Fourier spectra, such material functions are defined. It should be noted, however, that these material functions will prove useful provided they give insight into material behavior and help in identifying the microscopic mechanisms involved. Therefore, physical interpretations of these material functions are still being proposed and applied to different complex fluids to examine their utility.

4.2.4 Experimental Issues

In general, oscillatory rheological measurements are difficult at very low and high frequencies. At very low frequencies, with lower deformation and strain rates, signal-to-noise ratio is low due the sensitivity of sensors (or least count). At higher frequencies, inertia effects are significant, while the rheometric analysis assumes absence of these. Mechanical and electrical disturbances during the operation of the rheometer can also lead to errors, which may be significant during analysis of higher harmonics in LAOS [26].

The presence of higher harmonics during oscillatory shear is not always due to terminal nonlinear response of the materials. One has to be careful about contributions from sample inertia [5]. It has been shown that during startup of oscillatory flow (during the time before the terminal steady state is reached), higher harmonics are observed. Therefore, it is essential to establish the terminal periodic response before analyzing the LAOS data [8].

Due to small gaps of rotational rheometer geometries, it is assumed that flow is one dimensional with tangential velocities. However, at large amplitudes and at the edges of the geometries, secondary flows can develop. These may also lead to seeming deviation from linear viscoelastic response or the appearance of higher harmonics. A homogeneous flow is also assumed (with constant velocity gradient, or strain rate) across the depth in the rheometer geometry. During LAOS, the flow may become inhomogeneous (shear banding; regions of different strain rates) and therefore, analysis of steady terminal response would be very difficult [17].

Even though the presence of even harmonics is ruled out due to material symmetry considerations, they have been observed in several cases [4, 7]. These even-harmonic terms can be observed due to transient responses, secondary flows, viscous heating or dynamic wall slip. As an example, they have been shown to arise due to misalignment of top and bottom geometries [4]. Edge fracture has also been shown to lead to higher harmonics including even ones [17].

The intensity of $(n + 1)$th harmonic is usually in the range of 1–10% of the nth harmonic. Therefore, intensities of successive harmonics are very low and it is difficult to ascertain their significance. Due to these experimental issues, LAOS data need to be examined very closely before physical interpretation of material response.

4.3 Constitutive Models and LAOS

The linear viscoelastic models such as Maxwell, Jeffrey's, standard linear solid and generalized Maxwell model (combination of Maxwell model as described in Sect. 4.1.3) are commonly used to describe the oscillatory shear response [2, 16]. Some of the simplistic phenomenological models for large deformations or nonlinear viscoelasticity can be considered as extensions of the linear viscoelastic models. For example, upper convected Maxwell model is

$$\sigma + \lambda \left(\frac{\partial \sigma}{\partial t} + \mathbf{v} \cdot \nabla \sigma - (\nabla \mathbf{v})^{\mathbf{T}} \cdot \sigma - \sigma \cdot \nabla \mathbf{v} \right) = \eta \dot{\gamma}, \qquad (4.24)$$

where the partial derivative from (4.4) has been replaced with upper convected derivative (given in the parenthesis in the above equation).

The quasi-linear model such as upper convected Maxwell model or Lodge rubber-like liquid and Oldroyd-B model lead to linear shear stress response for SAOS as well as LAOS [5, 21]. Therefore, they predict single harmonic stress response, at the same frequency as the strain. However, normal stress differences are predicted to exhibit second harmonic [5] even at small strains. In addition, all these models predict the presence of higher harmonics during the initial stages of LAOS (before the terminal periodicity is established). These results are not surprising since the quasi-linear models exhibit no shear thinning behavior. As is apparent from cyclic response in Figs. 4.1 and 4.7, nonlinear viscous and nonlinear elastic behavior are required to observe hardening/softening during a cycle.

In terms of molecular/microscopic theories, these models can be considered to have *fixed entanglement network*. Models which consider the transitory nature of entanglement network are, therefore, essential to capture the LAOS. To understand the quantitative variation of stress, several molecular-based models and phenomenological models are being used to analyze the LAOS.

It has been shown that many of the prevalent constitutive equations show the following scaling for intensity of nth harmonic with strain [19, 29]:

$$I_n \propto \gamma_0^n. \tag{4.25}$$

The models such as Giesekus, Phan Thien Tanner, Finitely Extensible Nonlinear Elastic-P, Marrucci, Leonov, and Pom-Pom model show the behavior given by the above equation [7, 19]. The degree of shear thinning can be enhanced or reduced based on the parameters of these models. Attempts are being made to understand the LAOS response for different values of these parameters. In various experimental observations and simulations with the above models at very large amplitudes (>100%), departures from (4.25) have been reported.

Some of these studies highlight that models which are good for nonlinear viscoelastic behavior of complex fluids (exhibiting shear thinning, strain hardening in extensional flow, stress growth etc.) may not yield quantitative comparison with experimental observations for LAOS. Progress is needed on both the factors of limited understanding of the response of microscopic mechanisms to LAOS and limited interpretations of experimental measurements.

4.4 LAOS: Examples

The following list provides a few examples of material systems and specific issues that have been examined with LAOS:

- Examine morphology of polymer blend [4], probing of dispersed phase size distribution
- Effect of flow on drop shape in emulsions of Newtonian fluids [28]
- Concentrated particulate suspension [20]

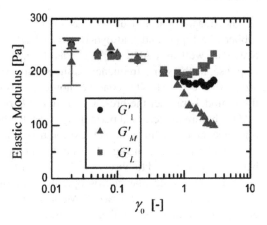

Fig. 4.10 Elastic moduli as a function of strain amplitude for mucus [7]

- Effect of long chain branching on oscillatory shear behavior [8]
- Effect of oscillatory shear on entanglement dynamics in polymers [22]
- Strain rate frequency superposition in soft solids [15].

Two examples, based on material functions defined in (4.22) and (4.23), are given in the following discussion.

Material functions for mucus (Lissajous plots shown in Fig. 4.7) are shown in Fig. 4.10 [7]. The response of the first harmonic shows in significant change (a slight decrease) with strain amplitude, while moduli defined in (4.22) show different response. The first observation would lead to a conclusion that near linear viscoelastic behavior is observed for almost the whole range of amplitudes investigated. Qualitatively, this behavior can be described as slightly softening behavior. However, based on G'_L and G''_M, one can observe that softening/hardening behavior is quite significant and this was also apparent from Lissajous plots.

Figure 4.11 shows LAOS measurements on linear and branched polymer melts [14]. The ratios of third harmonic to first harmonic ($Q\gamma_0^2$), as defined in (4.23), were measured for several linear and branched polymers and their blends. Based on simulations with several models (as shown in (4.25)), we expect this ratio to scale as γ_0^2. Indeed, for many polymeric melts and other materials, this has been observed. However, as the figure shows, presence of long chain branching leads to a slope <2. Differing response of linear and branched polymers to extensional flow is discussed in (Fig. 8.10). Based on the measurements shown in Fig. 4.11, LAOS can be said a useful tool to assess the effect of branching on rheological response of a polymer melt [14]. Pom-Pom model, developed to describe the behavior of branched polymers, also does not capture the departure of scaling from 2. Therefore, such results also point to the challenges in constitutive modelling of complex fluids under LAOS.

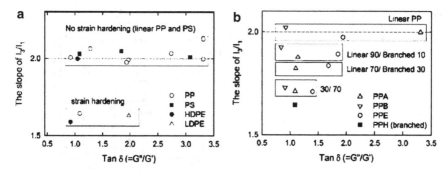

Fig. 4.11 Scaling of I_3/I_1 with strain amplitude for different (**a**) linear and branched polymer melts and (**b**) their blends [14]

4.4.1 Bulk Oscillatory Flow

In this section, an example of bulk oscillatory flow is presented. Figure 4.12 shows a cavity (a rectangular box) filled with a given fluid. The top surface of this cavity is moved periodically in x-direction. The filled fluid, at each point in the cavity, is also expected to have periodic velocity. This cavity was used to examine flow of different Newtonian and viscoelastic fluids.

A reactive flow, the gelation of sodium acrylate, was carried out in this cuboidal cavity with the top surface undergoing sinusoidal periodic motion. The instantaneous two-dimensional planar velocity fields during gelation were obtained using Particle Image Velocimetry [11]. Initially, the reaction mixture consists of monomer only, and therefore behaves as Newtonian fluid. However, as crosslinking reactions progress, viscoelastic behavior is developed (as shown in Fig. 4.2c). The periodic variation of velocities at different points was further examined by obtaining Fourier transforms of the time series of the velocities. Fourier transforms of x- and y-components of velocities at a point are shown in Fig. 4.13. These Fourier transforms were collected at different times after introduction of the initiator. These results show that point velocities fluctuate at higher harmonics during the time of reaction. It is interesting to note that such higher harmonics were observed neither with Newtonian glycerol/water mixtures nor with viscoelastic polyacrylamide/water mixtures.

Nonlinear response arising out of structural changes has been highlighted in the context of LAOS by showing appearance of higher harmonics during crystallization in a polymer melt [6]. It will be interesting to examine whether such behavior is observed during LAOS experiments during gelation.

Fig. 4.12 Periodically driven cavity to observe bulk oscillatory flow

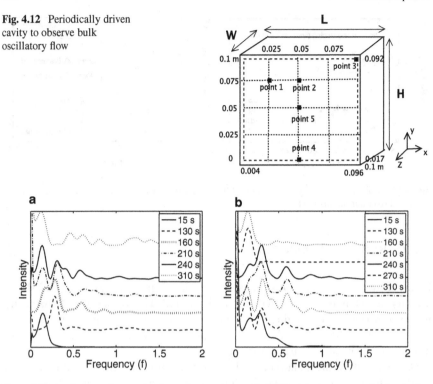

Fig. 4.13 Fourier transform of point velocity time series [11] (**a**) x-component of velocity (**b**) y-component of velocity

4.5 Summary

Nonlinear response of complex fluids is very important for most applications. In the last couple of decades, nonlinear response is being examined through various approaches. Oscillatory rheology at large amplitudes has been shown to be useful in analyzing nonlinear response of various complex fluids, such as polymers, emulsions, suspensions and gels. In this chapter, a brief description of LAOS was provided. Due to advances in instrumentation and experimental tools, it is possible to obtain large amount of data during LAOS. Through analysis of cyclic variation and frequency spectrum, the response can be analyzed to understand the material behavior. Constitutive models are also being used to understand material response under LAOS.

It has been shown that LAOS can highlight the structural changes taking place in the material at large deformations. The quantification of the nonlinear response, and more importantly, the physical interpretation, are still developing tools.

Acknowledgements The author thanks Subhash Naik, Padmarekha A, Sudharshan Parthasarathy, Moumita Sarkar and Manish Gupta for their help with experiments and figures.

References

1. Baumgaertel M, De Rosa ME, Machado J et al (1992) The relaxation time spectrum of nearly monodisperse polybutadiene melts. Rheol Acta 31:75–82
2. Bird RB, Hassager O (1987) Dynamics of polymeric liquids. Wiley, NY
3. Blythe T, Bloor D (2005) Electrical properties of polymers. Cambridge University Press, London
4. Carotenuto C, Grosso M, Maffettone PL (2008) Fourier transform rheology of dilute immiscible polymer blends: A novel procedure to probe blend morphology. Macromol 41:4492–4500
5. Collyer AA, Clegg DW (eds) (1998) Rheological measurement, 2nd edn. Chapman & Hall, London
6. Dotsch T, Pollard M, Wilhelm M (2003) Kinetics of isothermal crystallization in isotactic polypropylene monitored with rheology and Fourier-transform rheology. J Phys Condens Mat S923–S931
7. Ewoldt RH, Hosoi AE, McKinley GH (2008) New measures for characterizing nonlinear viscoelasticity in large amplitude oscillatory shear. J Rheol 526:1427–1458
8. Fleury G, Schlatter G, Muller R (2004) Non linear rheology for long chain branching characterization, comparison of two methodologies: Fourier transform rheology and relaxation. Rheol Acta 44:174–187
9. Guerrero A, Partal P, Gallegos C (2008) Linear viscoelastic properties of sucrose ester-stabilized oil-in-water emulsions. J Rheol 42(6):1375–1388
10. Harini M, Deshpande AP (2009) Rheology of poly (sodium acrylate) hydrogels during cross-linking with and without cellulose micro-fibrils. J Rheol 53(1):31–47
11. Harini M, Sriram S, Deshpande AP et al (2008) Variation of spatial and temporal characteristics of reactive flow in a periodically driven cavity: Gelation of sodium acrylate. Phys Rev E 78:1–8
12. Hyun K, Wilhelm M (2009) Establishing a new mechanical nonlinear coefficient Q from FT-Rheology: First investigation of entangled linear and comb polymer model systems. Macromol 42:411–422
13. Hyun K, Kim SH, Ahn KH et al (2002) Large amplitude oscillatory shear as a way to classify the complex fluids. J Non-Newt Fluid Mech 107:51–65
14. Hyun K, Baik ES, Ahn KH et al (2007) Fourier-transform rheology under medium amplitude oscillatory shear for linear and branched polymer melts. J Rheol 51(6):1319–1342
15. Kalelkar C, Lele A, Kamble S (2009) Strain rate frequency superposition in large amplitude oscillatory shear. arXiv:0910.0591
16. Larson RG (1998) Structure and rheology of complex fluids. Oxford University Press, NY
17. Li X, Wang S-Q, Wang X (2009) Nonlinearity in large amplitude oscillatory shear (LAOS) of different viscoelastic materials. J Rheol 53(5):1255–1274
18. Malkin AY (1995) Non-linearity in rheology an essay of classification. Rheol Acta 34:27–39
19. Nam JG, Hyun K, Ahn KH et al (2008) Prediction of normal stresses under large amplitude oscillatory shear flow. J Non-Newt Fluid Mech 150:1–10
20. Narumi T, See H, Suzuki A et al (2005) Response of concentrated suspensions under large amplitude oscillatory shear flow. J Rheol 49(1):71–85
21. Prost-Domasky SA, Khomami B (1996) A note on start-up and large amplitude oscillatory shear flow of multimode viscoelastic fluids. Rheol Acta 35:211–224
22. Ravindranath S, Wang S-Q (2008) Large amplitude oscillatory shear behavior of entangled polymer solutions: Particle tracking velocimetric investigation. J Rheol 52(2):341–358
23. Shih W-H, Shih WY, Kim S-I et al (1990) Scaling behavior of the elastic properties of colloidal gels. Phy Rev A 42(1):4772–4779
24. Thibierge C, L'Hote D, Ladieu F et al (2008) A method for measuring the nonlinear response in dielectric spectroscopy through third harmonics detection. Rev Sci Instr 79:103905–103910
25. Wilhelm M, Maring D, Spiess H-W (1998) Fourier-transform rheology. Rheol Acta 37:399–405

26. Wilhelm M, Reinheimer P, Ortseifer M (1999) High sensitivity Fourier-transform rheology. Rheol Acta 38:349–356
27. Winter HH, Mours M (1997) Rheology of polymers near liquid–solid transition. Adv Polym Sci 134:165–234
28. Yu W, Bousmina M, Grmela M et al (2002) Modeling of oscillatory shear flow of emulsions under small and large deformation fields. J Rheol 46(6):1401–1418
29. Yu W, Wang P, Zhou C (2008) General stress decomposition in nonlinear oscillatory shear flow. J Rheol 53(1):215–238

Chapter 5
PIV Techniques in Experimental Measurement of Two Phase (Gas–Liquid) Systems

Basheer Ashraf Ali and Subramaniam Pushpavanam

Abstract Flow visualization techniques help us understand qualitatively and quantitatively flow behavior in systems. The system walls have to be optically transparent for visualization. These methods require a good knowledge of optics, hydrodynamics, mathematics as well as electronics. Particle image velocimetry (PIV) is used to understand the quantitative spatiotemporal variations a the flow. It can give excellent spatial as well as temporal resolution. In this chapter, we discuss the basics of flow visualization from a theoretical and a practical view point and present results of PIV on a two-phase gas–liquid flow system.

5.1 Introduction

One of the early experiments based on flow visualization is the one carried out by Reynolds. In this classical experiment, he studied the flow through circular tubes by injecting a colored dye through a needle at different flow rates of the fluid. He observed that for low fluid velocities such that $Re < 2,100$, the dye injected from a needle moves along a straight line as a jet. Here, Re is the dimensionless Reynolds number, which is a ratio of inertial to viscous forces. For $Re > 2,100$, the jet or dye breaks up, confirming that the flow is turbulent. The experiment qualitatively describes the nature of the flow and helps understand the point of transition from the laminar to turbulent regime. The material used for pipe was transparent to allow visualization.

One of the necessary conditions for flow visualization is that the experimental setup must be transparent. Quantitative information of the flow can be obtained when we use a rotameter or a venturimeter. However, only a gross property of the flow i.e., the average velocity can be obtained in these devices. These techniques also are intrusive by nature. Here, the probe modifies the flow and the measured variable is modified by the intrusion of the used probe. Specifically, the probe may

S. Pushpavanam (✉)
Professor, Indian Institute of Technology Madras, Chennai 600036, India
e-mail: spush@iitm.ac.in

J.M. Krishnan et al. (eds.), *Rheology of Complex Fluids*,
DOI 10.1007/978-1-4419-6494-6_5, © Springer Science+Business Media, LLC 2010

induce an additional resistance and if this were to be significant, the flow through the system would be altered (decreased) drastically. We would then be measuring a value different from the one we intended to measure originally.

It is, hence, necessary to develop non-intrusive methods, which can give local quantitative information using a transparent set-up. By "local" we mean at a point in space or a region in space where the spatial variation can be determined. By measuring at different points in space, we can obtain the spatial distribution of the velocity (for example, the parabolic profile of laminar flow under steady fully developed conditions). Two techniques used for obtaining quantitative information of velocity are (i) LDV – Laser doppler velocimetry and (ii) PIV – Particle image velocimetry. LDV helps in obtaining quantitative information about velocity at a point. Here, we obtain temporal information about velocity at a point. Accurate information on turbulence characteristics at a point can be determined using LDV. However, no information on spatial variation is possible. For obtaining spatial information, the LDV measurement has to be carried out at different points of interest, i.e., we have to scan the region. This can be used when the flow is steady. Here, two laser beams are used and the fluid is seeded with particles. The measurement is based on the Doppler Effect.

PIV, on the other hand, measures velocity data i.e., flow-field in a plane. Here, we obtain both spatial and temporal information on the velocity field. Here, we compromise on temporal resolution as compared to LDV but obtain spatial information. These two methods are based on lasers, LDV on laser beams and PIV on laser sheets. These methods need a good understanding of the fundamentals of optics and image processing in addition to hydrodynamics.

5.2 Fundamentals of Optics

Jellyfish in an ocean are transparent and this helps them to catch their prey. They can be detected by the shadow they cast. This arises due to the difference in refractive index between the surrounding water and the jellyfish. The twinkling of stars is attributed to variations in the refractive index induced by changes in temperature and pressure in the upper atmosphere. This causes the light to deviate from the straight path and makes stars twinkle. Changes in refractive index have to be factored in when analyzing systems using flow visualization. These could arise due to the mismatch in the refractive index of the fluid and the container (glass or Perspex).

5.2.1 Scattering of Light

Light rays move in a straight line. When they encounter a particle, light is reflected or scattered. There are two modes of scattering. Mie Scattering for large sized particles when dp $>$ λ. Here, all wavelengths are scattered equally. The color of clouds in the sky is white since water drops have dp $>$ λ. Here, the incident white light is scattered uniformly. Rayleigh scattering occurs when dp $<$ $(\lambda/10)$. Here, scattering is inversely proportional to λ^4. White light is composed of **V**iolet, **I**ndigo, **B**lue, **G**reen, **Y**ellow, **O**range and **R**ed (VIBGYOR). Blue light is scattered more than red since it has a smaller wavelength. The color of the sky is blue due to Rayleigh scattering by small dust particles in the atmosphere. To avoid this behavior of preferential scattering of a wavelength as compared to other wavelengths, PIV employs lasers where the incident light is of a fixed wavelength.

5.2.2 Aliasing

Whenever a continuous signal is measured at discrete time instants, aliasing can occur. Two different continuous signals may have the same discrete samples as shown in Fig. 5.1. Thus from the discrete samples, it will be impossible to determine the original signal which contributed to the samples $(+)$. This is called aliasing, and a straightforward approach to overcome this to increase the sampling frequency. Another artifact of aliasing is now discussed. Consider a wheel with a mark $(\star)$ on it. Assume the time period of revolution is 60 s. If we measure the position of the mark every 20 s, the mark is observed at positions 1, 2, 3 cyclically and the wheel appears to move in a clockwise direction (the true direction). These positions are repeated (Fig. 5.2a). If we measure the position of the mark every 40 s, the position of the mark will appear at 1', 2', 3' and the wheel would appear to move counterclockwise (Fig. 5.2b).

As another example, consider the earth moving around the sun every 24 h. If the position of the sun is measured every 23 h, the motion would appear from west to east, which is opposite of the actual motion. A real signal is composed of a mixture of time periods or frequencies. Here, aliasing can be avoided by sampling at a time interval, which is less than half of the lowest period. These effects imply that we

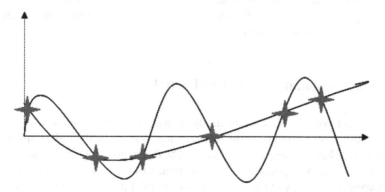

Fig. 5.1 Two continuous signals having the same discrete samples

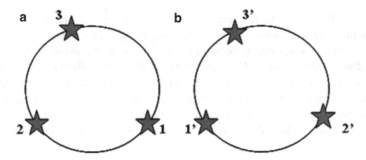

Fig. 5.2 Location of a mark on a rotating wheel with period of revolution 60 s measured every (**a**) 20 s (**b**) 40 s

have to measure the signal at a sampling period less than one half of the lowest time period of the signal to obtain an accurate representation of the signal.

5.2.3 Optical Filters

Different types of optical filters can be used to characterize single-phase and two-phase flows. These include band pass filters, long wavelength pass filters, small wavelength pass filters and fluorescent filters. White light is made up of seven colors. Each color has its own wavelength and the light has a continuous spectrum of wavelengths. Filters allow a selected range of wavelengths to pass through for detection. For a green (532 nm) band pass filter, the output has a narrow Gaussian shape with the mean at 532 nm where the intensity is a maximum. If the light source is an Nd:Yag laser, it emits light at a wavelength of 532 nm.

A long wavelength pass filter is used to cut off the wavelengths below a certain wavelength (if the filter has cut-off wavelength 575, it allows components more than 575 nm to pass through). This filter is useful for two-phase flow experiments where it is very important to differentiate the light emitted by the particles from the light emitted by the bubbles. The filter used is determined by the nature of particles as well as the nature of light source.

5.3 Fluid Mechanics of Tracer Particles

In PIV, tracer particles that are neutrally buoyant (have the same density as liquid) are added to the fluid. These particles are typically hollow glass spheres or polystyrene spheres of size around 10 μm. The difference in refractive index helps us to obtain images of particles. PIV is an indirect measurement technique. Here, we measure the velocity of the seed particles and infer the velocity of the fluid from it. We assume that the particles follow the liquid faithfully.

Following Kompenhans et al. [6], the velocity of the particle in a continuously accelerating fluid with acceleration "a" is given as

$$U_p - U_{fluid} = \left(\frac{d_p^2(\rho_p - \rho_f)}{18\mu} \right) a, \tag{5.1}$$

where U is the velocity, ρ the density, d the diameter and μ the viscosity. Here, the subscripts 'p' and 'f' are used to denote the particle and fluid phases.

The relaxation time i.e., the time required for the particle to attain the fluid velocity or to respond to changes in fluid velocity is given by

$$\tau = \frac{d_p^2(\rho_p - \rho_f)}{18\mu}. \tag{5.2}$$

It can be seen that the particle size must be small for a low relaxation time or for the particle to follow the liquid instantaneously. The size of the particles hence must be optimal. If the size is big, it may not follow the liquid. If the size is too small, then the light scattered by it would be very low and the images of particles would not be distinct and Brownian motion effects would dominate.

From (5.1), it is seen that the diameter of the particle should be very small to ensure accurate tracking of fluid velocity. Since light scattered from the particles is necessary to determine particle images and calculate flow field, the compromise in size has to be made.

5.4 Principle of Particle Image Velocimetry

The ability to quantify the instantaneous flow field of systems has been realized recently through the use of PIV. PIV is a non-intrusive technique that allows the measurement of transient flow trends. Thus, in PIV we can assess the coupling between velocities at different points of the flow field. Spatial correlations in the velocity time series can be obtained which is not possible using point measurement techniques, such as LDV. The primary element of the PIV system consists of a pulsed laser source and corresponding optics to produce a laser sheet. A pulsed laser source allows generation of high power to get good resolution of images. Since we need information on an area in space, the laser beam is converted to a sheet. The images of the particles are determined at two time instants using CCD cameras. These are transferred to a computer for processing and obtaining velocity fields.

PIV is based on determining the movement of a group of particles seeded into the flow. The area of interest is illuminated by two consecutive short duration light pulses produced by a laser (e.g., double-pulse Nd:Yag). The duration of the light pulses and the time interval (Δt) between two pulses is typically of the order of milliseconds (ms) and this is decided by the magnitude of velocities prevailing in the region. The light scattered from the seed particles is captured using charge coupled device (CCD) cameras (typical resolution: $1,280 \times 1,024$ pixels) and stored for subsequent analysis. The time interval (Δt) between two pulses also depends on the camera, the acquisition mode and the type of laser. The value of Δt is limited by the time necessary for transfer of frames from the camera to the computer. The velocity of the particles can be estimated by finding the displacement of the particles between the laser pulses using

$$u = \frac{\Delta x}{M \Delta t}, \tag{5.3}$$

where M is the magnification of the camera. The displacement of the particles between two images can be determined by using a cross correlation technique, which we will elaborate on later.

Since the time interval between the laser pulses is known and the camera is calibrated to yield the displacement in the object plane (in meters), the flow velocity can be determined. In practice, the entire image is divided into small sub-regions called interrogation areas, and the local displacement of the particles is determined within each interrogation area. In each interrogation window, all the particles are assumed to have the same velocity. The cross-correlation technique determines the displacement which best matches the images of the particles in the first frame or image to those in the second frame. Increasing the number of interrogation windows results in a higher resolution.

When this is done for all interrogation areas of the image, it results in an instantaneous velocity field for the entire cross-section of the flow. The main difference between LDA and PIV is that the LDA is a single point measurement technique with a high temporal resolution, whereas PIV is a whole plane measurement technique, with a high spatial resolution. Here, from one double-frame image, several thousand velocity vectors are computed, which are arranged in a grid.

5.4.1 Components of Particle Image Velocimetry

Now we will discuss the basic components of PIV as applied to two-phase gas–liquid systems. The technique uses particles and their images to determine the velocity fields.

Laser: The Nd:Yag laser commonly used in PIV system generates a visible green laser (532 nm). The pulsation frequency as well as the energy level must be determined depending on the flow system being analyzed. Laser is an acronym for **L**ight **A**mplification by **S**timulated **E**mission of light **R**adiation. Lasers are made of

- Laser material [He–Ne (red light), Nd–Yag (green light)]
- Pump-flash lamp or external source of energy
- Optical cavity or mirror arrangement

We first discuss the case when the atoms of the laser material can exist in two states E_u, E_l (the upper and lower levels), such that $E_u - E_l = \mathbf{h}\gamma$. When the atoms in E_l absorb photons with $\mathbf{h}\gamma_l$, they go to the higher energy state E_u and when atoms in E_u emit photons with energy $\mathbf{h}\gamma_1$, they relax to energy state E_1. Consider an atom in E_u when one photon impinges on it with $\mathbf{h}\gamma_1$. The atom can relax to E_1 and we obtain two photons each with energy $\mathbf{h}\gamma_1$. These two photons can give rise to four photons on impinging on atoms in E_u. Hence, in addition to the material which needs to exist in two states of energy, every laser must have an external energy source called "pump." This increases the number of atoms at the upper energy level E_u. The aim is to increase the number of atoms in E_u as compared to E_1. This is called population inversion.

Lasers with three energy levels also exist. Here, there are three levels E_u, E_1 and E_3. When photons with energy $\mathbf{h}\gamma_3$ are injected, atoms will be at energy level E_3 $(> E_u)$. These instantaneously relax to E_u. The atoms spend some time in E_u and relax from E_u to E_1, emitting photons of energy "$\mathbf{h}\gamma$." This gives the lasing action. The energy pump (which could be a flash lamp) ensures that more atoms are present in E_u. These atoms relax emitting photons with energy "$\mathbf{h}\gamma$." These photons are reflected by the mirror (Fig. 5.3c) and they impinge on atoms at E_u. This generates more photons releasing energy. A small fraction of the photons is emitted and comes out as a laser beam with a fixed wavelength and the remainder is used to generate more photons to sustain the lasing action.

Cameras: The cameras used in PIV system are based on an interline CCD sensor. These have high resolution and high sensitivity. The images are stored as intensity values in each pixel and this is used as data in the cross correlation for velocity determination.

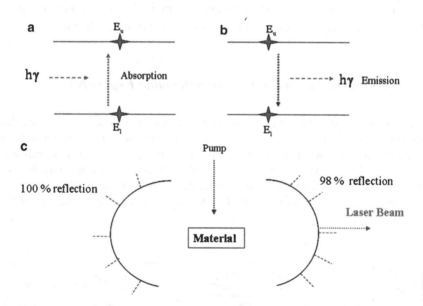

Fig. 5.3 (a) Absorption of photons (b) emission of photons (c) mirror or optical cavity arrangement

Fluorescence represents the ability of a material to absorb light at a particular wavelength and emit light at a higher wavelength. A fluorescent molecule has an absorption spectrum, i.e., it absorbs light at a fixed wavelength. For instance, Rhodamine (a carcinogenic dye) absorbs light at $\lambda = 532$ nm and emits red light ($\lambda = 564$ nm). The energy of the photon absorbed is $h\gamma_a$, while the energy released is $h\gamma_r$. $h\gamma_r < h\gamma_a$ as the energy released has to be less than the energy absorbed. This results in $\gamma_r < \gamma_a$ or $\lambda_r > \lambda_a$. Consequently, the wavelength of the light emitted is always higher than the wavelength of incident light. Ultraviolet (uv) light is usually used as the source of light since its wavelength is low and any light emitted by fluorescent molecules will have a higher wavelength and be in the visible spectrum.

5.4.2 Auto- and Cross-Correlations

The objective of the PIV measurements is to statistically determine the displacement between two patterns of the particle images. These images contain intensity information stored as a 2D distribution. It is common practice to determine the shift in time between two almost identical signals using correlation techniques. The theory of correlation techniques in PIV is a straightforward extension from the 1D time signal case to the 2D spatial distribution case.

The autocorrelation function can be determined in two ways: a direct numerical simulation or using the Wiener–Kinchin theorem. This theorem exploits the fact that the Fourier transform of the autocorrelation function R1 and the power spectrum $|\hat{I}(x, y)|^2$ of an intensity field $I(x, y)$ are Fourier transforms of each other.

5.4.3 Autocorrelation (Single Frame/Double Exposure)

Here, the scattered light from both the first and second exposures of the particles is recorded in one image. The complete image is subdivided into smaller regions called interrogation windows and the velocity vector in each window is evaluated by autocorrelation (Fig. 5.4). The autocorrelation function is characterized by two identical correlation peaks located symmetrically about the highest central peak indicating zero displacement [3, 7, 9]. Due to this, we cannot detect the sign of the

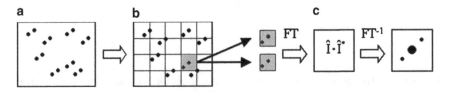

Fig. 5.4 (a) Image of particles (b) interrogation windows (c) computation of displacement

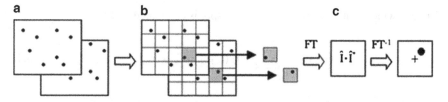

Fig. 5.5 (**a**) Image of particles (**b**) interrogation windows (**c**) computation of displacement

displacement because prior information about which particle is illuminated by the first and the second laser pulses is lost. So the information from the autocorrelation is ambiguous and not conclusive unless we use some a priori information about the observed flow. Also, the detection of very small displacements is a problem as in this case the correlation peaks are very close to the central peak. The calculation of the autocorrelation function using Fourier transforms is shown in Fig. 5.5.

5.4.4 Cross-Correlation (Double Frame/Double Exposure)

The scattered lights from first and second exposures of the particles in this approach are recorded into two different images. The complete image is subdivided in interrogation windows and the velocity in each window is evaluated by cross-correlation.

Compared to the autocorrelation, a higher and unambiguous correlation peak is obtained. The autocorrelation peaks are significantly smaller than the cross-correlation peaks. High noise increases the possibility that the displacement correlation peak disappears in the background and the accuracy of the autocorrelation technique is reduced [3, 7, 9].

5.4.5 Correlation Averaging Function

The critical step in the PIV is to get accurate images of the particles. For this, the laser sheet and the camera must be synchronized. After obtaining the images, the velocity of the particles must be determined. The relationship $U = \frac{\Delta s}{\Delta t}$ is used where Δt is the time gap between two laser pulses. Δs is the displacement, which needs to be found out.

While using cross correlation, the images are stored in two frames. The saved images contain intensity values at each pixel. A typical camera has $1{,}280 \times 1{,}024$ pixels. Each image is divided into interrogation windows of size 32×32 pixels. This results in 40 vectors in one direction and 32 in another (a total of 1,280 vectors). The correlation functions between each window in the first image and the corresponding window in the second image are estimated. This generates the correlation function as a matrix. For correlation-based PIV evolution algorithms, the correlation function for each window at a certain interrogation spot is calculated by [2],

$$\varphi_k(\mathbf{m}, \mathbf{n}) = \sum_{j=1}^{q} \sum_{i=1}^{p} \mathbf{f}_k(\mathbf{i}, \mathbf{j}) \cdot \mathbf{g}_k(\mathbf{i} + \mathbf{m}, \mathbf{j} + \mathbf{n}), \qquad (5.4)$$

where $\mathbf{f}_k(\mathbf{i}, \mathbf{j})$ and $\mathbf{g}_k(\mathbf{i}, \mathbf{j})$ are the gray value distributions of the first and second exposures, respectively, in the kth PIV recording pair at a certain interrogation spot of size $p \times q$ pixels.

The correlation function for a singly exposed PIV image pair shows a significant peak at the position of the particle image displacement in the interrogation area. This dominates other peaks in the area. If there is too much noise in the images, the main peak becomes weak and may be lower than the other peaks. This can give rise to spurious vectors and an erroneous velocity vector is generated. This has to be handled in the post-processing stage. For steady-state flow conditions, the velocity field is independent of measurement time. Hence, the main peak $\phi_k(m, n)$ is always at the same position for PIV recording pairs taken at different time instants. Here, the side peaks appear with random intensities and positions in different recording pairs.

Therefore, when averaging ϕ_k over a large number of PIV recording pairs, the main peak will remain at the same position in each correlation function but the other peaks, which occur randomly, will average to zero. The ensemble averaged correlation function, as implemented in many PIV software tools, is given as [2]

$$\varphi_{\mathrm{ens}}(\mathbf{m}, \mathbf{n}) = \frac{1}{N} \sum_{k=1}^{N} \varphi_k(\mathbf{m}, \mathbf{n}). \qquad (5.5)$$

For each interrogation window, the maximum peak location and the corresponding displacement is determined. This can alternatively be determined using FFT.

5.4.6 Optimization of Particle Image Shift

For a given flow velocity and factor of magnification, the selected pulse delay dt determines the separation ds of the particle images on the CCD camera. The optimum separation of particle images depends on the desired interrogation window size d_{IntWin} and on the velocity gradients in the PIV recording. In general for cross-correlation, the separation of the particle images (in pixel) should be larger than the accuracy of the peak detection and smaller than a quarter of the selected interrogation window size (in pixel) [4, 5]:

$$0.1px < ds < 1/4d_{\mathrm{IntWin}}.$$

Additionally, the deviation of the particle separation Δds of all particle pairs within one interrogation window should be smaller than the mean particle diameter d_{p}:

$$\Delta ds < d_{\mathrm{p}}.$$

This imposes a limit on maximum interrogation window size and the sampling rate depending on the turbulence of the observed flow field.

5.4.7 Post Processing

The images taken by the cameras are post processed using software packages to obtain the velocity vector field. There may be some spurious vectors in the raw PIV velocity vector field due to noise, which arises during image acquisition and processing. We now discuss how this can be addressed. The spurious vectors are removed by setting an allowable vector range for the velocity components. This range is fixed after inspecting the different images. All the vectors outside this range are removed. The gaps created from the removal of vectors are filled by interpolation. Care must be taken to ensure that the post-processing operation does not tamper with the flow features of the velocity field. The images are post processed using a multipass technique where the size of interrogation window is progressively decreased starting from a larger size, say 128×128 pixels, and going up to a smaller size, say 32×32 pixels. This yields an instantaneous field of 32×40 vectors, i.e., a total of 1,280 vectors throughout the field (if we choose an interrogation window of size 32×32 for a camera having $1,024 \times 1,280$ pixels).

Figure 5.6 depicts two successive images obtained at a time "Δt" apart. The system consists of a rectangular tank with water. Gas is introduced at the lower right corner. The two frames i.e., the upper and lower parts contain the two images. The upper (lower) part of the image contains the first (second) exposure. On the right side of the frames, we see the gas bubbles, which are around 3 mm in diameter. The particles in the frames may not be clearly visible. They are identified using the in-

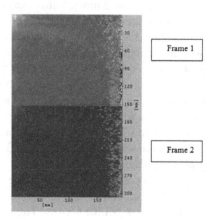

Fig. 5.6 Image of flow system

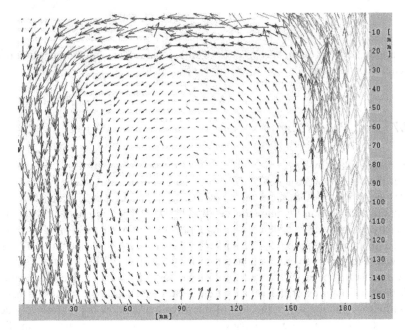

Fig. 5.7 Flow field for the raw image, interrogation window size 32 × 32 pixels

tensity values in the pixels. For the evaluation of PIV velocity vectors, the complete
image is divided into smaller square interrogation windows (32, 64, 128 pixels).

The algorithm computes the cross-correlation of each interrogation window from
the upper part (frame 1) with the corresponding window in the lower part of the image (frame 2). This yields the vector field shown in Fig. 5.7. It is seen that some of
the vectors are spurious (have an unsteady magnitude or different direction compared to their neighbors).

To address this, a median filter can be used. The median filter computes a median
vector from the eight neighbouring vectors. The final vector field can be smoothed
by a simple 3 × 3 smoothing filter to reduce noise. During this, the empty spaces
are filled up using two neighbour vectors through interpolation techniques. The final
post-processed vector is shown in Fig. 5.8.

To illustrate the effect of interrogation window on the flow field we depict the
flow field when the interrogation window size is 64 × 64. For example, with an
interrogation window size of 64 × 64 pixels, an image of e.g., $1,280 × 1,024$ pixels
is divided into 20 × 16 interrogation windows and it is depicted in Fig. 5.9. For an
interrogation window size set to 32 × 32 pixels, for the image of e.g., $1,280 × 1,024$
pixels, we obtain 40 × 32 vectors.

It is also possible to determine the velocity field using an overlap of the interrogation windows. For an interrogation window of 32 × 32 with a 50% overlap we
obtain 80 × 64 (5,120) vectors. This is shown in Fig. 5.10. The evaluation of PIV
image can be done using a smaller overlap value when the velocity field is reason-

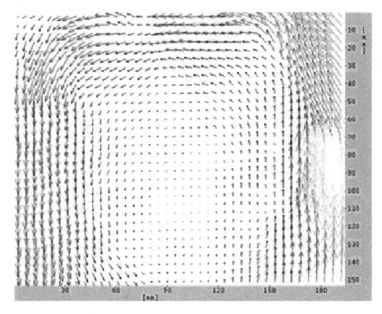

Fig. 5.8 Post processed instantaneous velocity field for the interrogation window size 32 × 32 (single pass with constant window size)

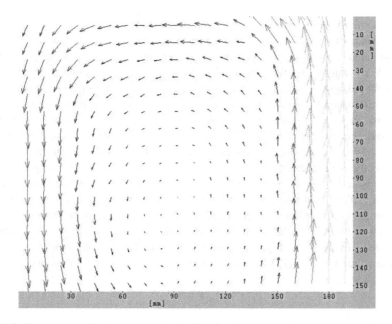

Fig. 5.9 Post-processed instantaneous velocity field for the interrogation window size 64 × 64

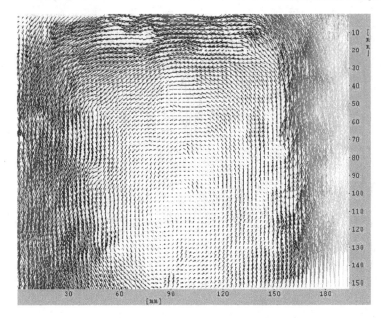

Fig. 5.10 Post-processed instantaneous velocity field for the interrogation window size 32 × 32 (50% overlap)

ably smooth. Now the effect of single pass processing (with constant window size) is compared with multi-pass processing (with decreasingly smaller window sizes) for the calculation of PIV image. The single pass uses a constant window size for the PIV evaluation and the flow field using this method is shown in Fig. 5.8. In the multi-pass, we decrease the process with a larger interrogation window of size 128 × 128 to smaller sizes. The evaluation of PIV image starts in the first pass with the initial interrogation window size and calculates a reference vector field. In the next pass, the window size is half the previous size and the vector calculated in the first pass is used as a best-choice window shift for the second pass. In this manner, the velocity vectors are calculated accurately and more reliably. This improves the spatial resolution of the vector field and produces less erroneous vectors. The calculation of flow field with this method is shown in Fig. 5.9. When one compares the flow field obtained by these methods, the single pass with constant window size does not capture recirculation of flow especially near the right top end (Fig. 5.8) compared to multi-pass (Fig. 5.11). So, we have used single pass with decreasing smaller window size for PIV evolution to enhance accuracy in the flow field.

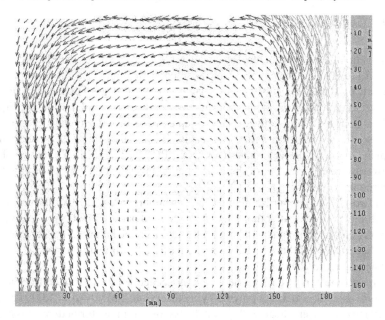

Fig. 5.11 Post-processed instantaneous velocity field with the help of filter (32 × 32)

5.5 Case Study

Most processes in the chemical process industry involve the flow and contact of multiple phases. One can, for example, think of the gas–particle flow in a fluidized bed or the gas–liquid flow in a bubble column. According to Tatterson [8] 25% of all chemical reactions occur between gas and liquid. A major class of gas–liquid flows is the one where the liquid phase is continuous and the gas phase is dispersed in the form of bubbles. In this work, this class is referred to as "gas–liquid flows." The rate-determining resistance plays an important role in the choice of reactor type for a certain process. When the reaction is the rate-determining step, bubble columns are often used, because of their large liquid bulk. Bubble columns are also used in the process of stripping or absorption. For both processes, the rate of transfer of components from the interface into the liquid phase is important. When the transfer of the component from the gas–liquid interface to the bulk of the liquid is rate determining, the reactor capacity can be increased through the use of a stirrer. To obtain an in-depth understanding of the performance of the reactor, detailed knowledge of the hydrodynamics is vital. The hydrodynamics of this reactor is studied through experiments and numerical simulations here [1].

5.5.1 Experimental

A schematic of the experimental set-up and main components of the PIV are shown in Fig. 5.12. Here, a rectangular tank is filled with liquid and air is introduced through a porous sparger in the lower right corner. The gas bubbles rise up due to buoyancy and drag the liquid with it. This induces a circulation in the liquid, which is measured quantitatively. The set-up consists of a rectangular tank (height 30 cm, width 20 cm, depth 2 cm). The results of measurement of the velocity field in a plane made in the rectangular tank using PIV for the gas flow rates of 21 pm are now discussed. The area of the experimental investigation where the flow field is determined is shown in Fig. 5.12.

In a two-phase system (gas–liquid), the gas bubbles are usually of the order of a few millimeter. They scatter more light than the seed particles and hence the images of the particles are masked by the light reflected from the bubbles. It is hence necessary to separate the light reflected from the particles with that reflected by bubbles to obtain the liquid velocity field.

To overcome this problem, the liquid is seeded with tiny, neutrally buoyant rhodamine-coated particles of polymethylmethacrylate, PMMA ($\rho = 1,000\,\mathrm{kg\,m^{-3}}$, size 10μm). A laser sheet with a thickness of 1 mm is formed by passing a

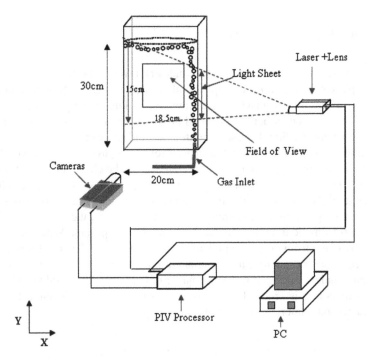

Fig. 5.12 Schematic diagram for PIV measurement in two-phase flow experiment (bubble column)

double-pulsed Nd–Yag (532 nm, 120 mJ) laser beam through an optical arrangement consisting of cylindrical and spherical lenses. The sheet illuminates the plane of interest. A CCD camera is positioned perpendicular to the plane of the light sheet to capture the light scattered from the rhodamine-coated particles. The emitted light from these particles is at a higher wavelength ($\lambda = 560$ nm, red) than the incident green light ($\lambda = 532$ nm), which allows only the red light to pass. An optical filter is placed in front of the camera which allows only the emitted red light from the fluorescent particles to enter the camera by filtering unwanted green light. The particles in the flow field are illuminated twice at a time interval of 1,400–2,100 μs by the laser sheet. The displacement of particles in the time between the laser pulses is recorded by capturing the image of the particles in each pulse. The displacement in the particle position in the image is obtained using a cross-correlation technique, which is numerically efficient.

The recorded particle displacement field measured across the whole field of view is scaled by the magnification of the camera and then divided by the pulse separation to obtain the velocity vector at each point. Scattered light from the first and second exposures of the particles is recorded in two different images. The algorithm computes the cross-correlation of all interrogation windows from the first frame with the corresponding window on the second frame.

For the evaluation of the liquid velocity field from the particle images, it is assumed that the tracer particles follow the local flow faithfully between two illuminations. So the particle velocity directly measures the liquid velocity. An optical filter is placed in front of the camera, which allows only the emitted red light from the fluorescent particles to enter the camera. The filter helps us to capture the reflection from the particles in the liquid by filtering out the unwanted green light reflected by the bubbles. This helps us to differentiate between the velocities of the two phases and only the liquid phase velocity is measured as only the liquid contains the fluorescent particles. Three hundred images were taken for each gas flow rate. These were processed to get the liquid velocity vector field in the entire plane.

5.5.2 Results and Discussion

As an example, we now discuss the flow induced in a liquid by injecting gas at a corner [9]. The long-time behaviour of x component of velocity is analysed at $z = 0.01$, the central plane (along the depth). The instantaneous planar velocity fields were time averaged and plotted in Fig. 5.13a. These are compared with the flow field obtained by simulations in Fig. 5.13b.

Figure 5.13 compares the time averaged velocity field observed experimentally (Fig. 5.13a) with the predictions of simulations as obtained from a commercial package FLUENT (Fig. 5.13b). We observe that the streamlines are in the form of closed curves in the counter clockwise direction and the flow field measured and predicted confirms this.

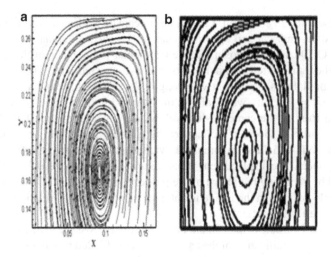

Fig. 5.13 Time averaged flow field in the field of view for a gas flow rate (**a**) experiment, (**b**) simulation

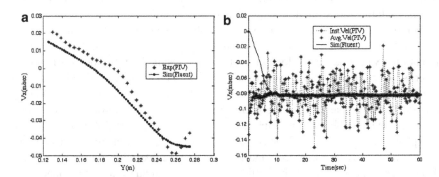

Fig. 5.14 (**a**) Time averaged velocity (Vx) along vertical line $X = 0.0121$ m, (**b**) Temporal variation of liquid velocity (Vx) at a point (0.025, 0.2486)

For the quantitative comparison of the experimentally determined velocity vectors with the model predictions, the time averaged x component of velocity plotted along the vertical line is shown in Fig. 5.14a. The time-dependent velocity components at a point (0.025, 0.2486) are also extracted from the instantaneous velocity fields to extract turbulent features of the flow (Fig. 5.14b).

Spatial variation of velocity component along vertical line obtained from PIV experiment agrees well with Computational Fluid Dynamics(CFD) simulations (see Fig. 5.14a). The fluctuating part in Fig. 5.14b indicates the experimentally measured instantaneous velocity at a point in the flow. This is time averaged and compared with the CFD simulations. We see that the time averaged predictions for both components compare favourably with the experimentally determined time averaged values. The constant value of the time averaged velocity indicates that the system does not show any oscillatory behaviour as the source of the gas is close to the wall.

5.6   Issues in Particle Image Velocimetry

- One of the frequently asked questions is how do we analyze particles in the second image which are not present in the first image and vice versa. This is usually addressed in the post-processing stage where we assume the displacement is periodic across the boundaries. There are other mathematical approaches to handle this issue.
- Calibration is an important step when quantitative information is desired. In PIV, a calibration sheet with markings in pre-determined equispaced intervals is immersed in the fluid medium being studied. The image of the sheet is taken by the camera and the distance in the object is correlated with images in the pixels. We determine how many pixels correspond to an actual distance in the object plane.
- To avoid the particles from the plane of interest leaving from the front or back, the time intervals between the pulses must be varied. If many particles leave from the front or back, Δt must be reduced. This also is determined by the velocity magnitude parameter.
- The laser pulse rate and camera frame rate must be synchronized.
- When the surface of the system is cylindrical, the curvature of the cylinder distorts the actual distances in the calibration sheet in the image. Different regions in the image plane would be affected through different magnifications and this has to be accounted for while obtaining quantitative measurements. One approach usually adopted is to immerse the system in a rectangular tank, which has a fluid with the same refractive index as the fluid used and the material of the system.

References

1. Ashraf Ali B, Siva Kumar Ch, Pushpavanam S (2008) Analysis of liquid circulation in a rectangular tank with a gas source at a corner. CEJ 144:442
2. Hessel V, Renken A, Schouten JC, Yoshida JI (2009) Handbook of Micro Process Technology, Volume 1: Fundamentals, operations and catalysts, Wiley-VCH Weinheim
3. Hinsch K (1993) PIV, speckle metrology, Marcel Dekker, NY 235–324
4. Keane RD, Adrian RJ (1990) Optimization of particle image velocimetries with mutiple pulsed systems. Paper 12.4, Proc 5th Intl Sym Applications of Laser Techniques to Fluid Mechanics
5. Keane RD, Adrian RJ (1990) Optimization of particle image velocimeters, Part I: Double pulsed systems. Meas Sci Tech 1:1202–1215
6. Kompenhans J, Raffle M, Willert (1998) Particle image velocimetry – A practical guide, Springer, Berlin
7. Lavison (2002) PIV hardware manual for Davis 6.2, Gottinen, Germany
8. Tatterson GB (1991) Fluid mixing and gas dispersion in agitated tanks, McGraw-Hill, NY
9. Westerweel J (1993) Digital PIV: Theory and applications, Delft University Press, Delft

Chapter 6
An Introduction to Hydrodynamic Stability

Anubhab Roy and Rama Govindarajan

Abstract In this chapter, our objective is twofold: (1) to describe common physical mechanisms which cause flows to become unstable, and (2) to introduce recent viewpoints on the subject. In the former, we present some well-known instabilities, and also discuss how surface tension and viscosity can act as both stabilisers and destabilisers. The field has gone through a somewhat large upheaval over the last two decades, with the understanding of algebraic growth of disturbances, and of absolute instability. In the latter part we touch upon these aspects.

6.1 Introduction

> *"... not every solution of the equations of motion, even if it is exact, can actually occur in Nature. The flows that occur in Nature must not only obey the equations of fluid dynamics, but also be stable."*

> – Landau & Lifshitz

Flow stability has preoccupied fluid dynamicists for centuries. The ubiquitous nature of turbulence, and the inability to offer a universal theory of it, led many to try tackling what was perceived as a simpler problem: of transition from a laminar state to a turbulent one. In the last two centuries, such efforts many-a-times have offered remarkable insight and helped make successful predictions. The description has undergone major reforms from time to time, and the complete process is not fully understood. The transition to turbulence begins usually with an instability of the laminar state, which is the subject of this brief review. An effort is made to address the basic tenets of hydrodynamic stability with a focus on a few recent viewpoints on the subject.

R. Govindarajan (✉)
Engineering Mechanics Unit, Jawaharlal Nehru Centre for Advanced Scientific Research, Jakkur, Bangalore 560064, India
e-mail: rama@jncasr.ac.in

J.M. Krishnan et al. (eds.), *Rheology of Complex Fluids*,
DOI 10.1007/978-1-4419-6494-6_6, © Springer Science+Business Media, LLC 2010

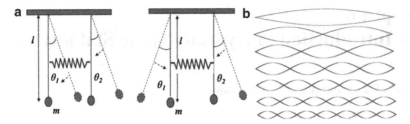

Fig. 6.1 Normal modes. (**a**) A double pendulum exhibiting 2 normal modes – symmetric and anti-symmetric oscillations (**b**) A vibrating string with the first seven of its infinite normal modes

Early studies in hydrodynamic stability arose in the theory of water waves by Newton followed by seminal work by Laplace, Cauchy, Poisson and later, by Stokes. Incidentally, most of the early work was done before the world got acquainted with Fourier series, although Cauchy employed defacto Fourier transforms [5, 6]. The advent of harmonic analysis facilitated the study of stability of mechanical systems by introducing the notion of normal modes. A normal mode is a pattern of oscillations in which the entire system oscillates in space/time with the same wavelength/frequency. A coupled pendulum with two degrees of freedom has two normal modes, whereas a vibrating string has infinitely many normal modes, often labelled as harmonics in musical instruments (Fig. 6.1). The beauty of normal modes lies in their ability to offer a complete description for the evolution of any arbitrary disturbance in *most* cases. Helmholtz, Kelvin and Rayleigh were among the earliest to use normal modes to describe problems in hydrodynamic stability. We will discuss some instabilities named after them.

6.2 Waves in Fluids

The variety of available restoring mechanisms allow a plethora of waves in fluids [20]. To name a few – compressibility gives rise to sound waves, variation in buoyancy leads to gravity waves, inertial waves owe their origin to rotation, magnetic fields lead to Alfven waves, whereas a latitudinal variation of Coriolis force leads to Rossby waves.

A typical example to understand the generation of oscillations in fluids is internal gravity waves. Consider a fluid at rest, whose density $\rho(z)$ increases with the depth z, i.e., the fluid is stably stratified, with heavier fluid at the bottom and lighter fluid above. A fluid parcel of density ρ_0 at z_0, when displaced upward by a distance δz, is now surrounded by lighter fluid and thus sinks. In the absence of retarding forces due to viscosity, it would then overshoot its mean position and reach a neighbourhood of heavier fluid, where it experiences an upward thrust (Fig. 6.2). From Newton's second law we have

$$\rho_0 \frac{d^2 \delta z}{dt^2} = -\delta \rho \, g, \tag{6.1}$$

Fig. 6.2 Oscillations in a
continuously and stably
stratified fluid

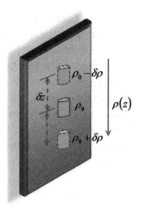

where g is the acceleration due to gravity and $\delta\rho$ is the density difference between the parcel and the ambient fluid at $z_0 + \delta z$. Since $\delta\rho$ is infinitesimal, it can be expressed as $\delta\rho = -\delta z\, d\rho/dz$. We may rewrite (6.1) as

$$\frac{d^2\delta z}{dt^2} + N^2\delta z = 0, \tag{6.2}$$

$$\text{where} \quad N = \sqrt{-\frac{g}{\rho}\frac{d\rho}{dz}}$$

is the oscillation frequency, called the Brunt–Väisälä frequency. Thus, buoyancy provides a restoring force in this oscillatory motion. The reader can easily predict the effect of viscous forces on this oscillatory motion.

Such wave-like motion in fluids is abundant in nature. Ripples generated on the surface of a pond to giant waves in oceans – all can be appreciated as oscillations over a quiescent flow state.

6.3 Instabilities

We have given an example of a stable system in the above section, where the oscillations are at best neutral, i.e., continue for ever with the same amplitude, and in most real situations, are damped out by a retarding force. What of the opposite situation, i.e., one in which small oscillations grow in amplitude? This is a common way in which a flow becomes unstable. In the above example, if the sign of $d\rho/dz$ were to be flipped, i.e., if the fluid were to be unstably stratified, one can see that the parcel of fluid would get accelerated away from its starting position rather than oscillate. In many flows, one may have growing oscillations instead, which could either be followed quite abruptly by a transition to turbulence, or be the first step in a long and eventful march towards turbulence. Flow instabilities occur all around us in both desirable and undesirable situations. Thanks to the efforts of numerous remarkable mathematicians, engineers and physicists, a lot is known about them, but more continues to bewilder us.

We consider a few model flow situations, concentrating on the physical mechanisms responsible for destabilising simple orderly flows and taking them to a completely different state. We first discuss linear instability, i.e., when the orderly state is given a very small perturbation, and examine whether the resulting oscillations grow or decay. For extensions to more complicated flows, the interested reader may consult standard textbooks, some of which are listed at the end of this chapter.

6.3.1 Effect of Shear: Kelvin–Helmholtz Instability

The classical problem serving as an excellent introduction to hydrodynamic instability is that of Kelvin–Helmholtz. Imagine two uniform streams of different velocities flowing past each other. Such a scenario is very commonly observed in nature, and is known as a mixing layer. The jump in the velocity across a layer of infinitesimal thickness is described as a vortex sheet (Fig. 6.3b). The vortex sheet consists of a constant density of point vortices, with their vorticities pointing out of the paper. The physical mechanism for the instability as described by Batchelor [2] relies purely on vortex dynamics. Let the vortex sheet be perturbed by a sinusoidal disturbance so that the perturbed interface is located at $\eta = \sin kx$. In the neighbourhood of the nodes (A and B), the positive vorticity induces a clockwise circulating velocity field. If $\partial\eta/\partial x > 0$, the crest and trough move away from each other, leading to vorticity being swept off from nodes like A, whereas if $\partial\eta/\partial x < 0$, the crest and trough come closer to each other, leading to vorticity being swept into nodes like B. Thus, accumulation of vorticity at points like B takes place unboundedly in the linear, non-dissipative scenario, giving exponential growth. This mechanism translates into the celebrated Rayleigh–Fjørtoft criterion in inviscid parallel shear flows – that there should exist a vorticity maximum somewhere in the base flow for instability to occur. To see the linear instability mathematically, we begin with the Euler equations in two dimensions (which are equivalent to the Navier–Stokes equations without viscosity). Here, u is the component of the velocity in the streamwise

Fig. 6.3 Kelvin–Helmholtz instability. (**a**) represents the unperturbed state and (**b**) the corresponding vortex sheet base state. A sinusoidal perturbation is imposed on the vortex sheet (**c**) which amplifies through vorticity induction (**d**)

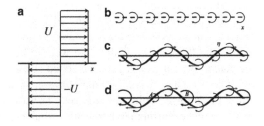

direction x, and v the component in the normal direction z. We then split all flow quantities into their mean and a perturbed value, such as $u = U(z) + \hat{u}(x, z, t)$. In two dimensions for an incompressible flow, the continuity equation $\nabla.\mathbf{u}$ enables us to write both components of the velocity vector $\mathbf{u}$ in terms of the stream function ψ, as $u = \partial\psi/\partial y$, and $-v = \partial\psi/\partial x$. The perturbed quantities are considered in their normal mode form, such as $\hat{\psi}(x, z, t) = \psi(z) \exp[ik(x - ct)]$, where k and c are the wavenumber and wave speed, respectively, with $c = \omega/k$, ω being the frequency of the disturbance. In the discussion here, we allow ω, and therefore c, to be a complex quantity, whose imaginary part tells us whether the disturbance grows or decays. We take the disturbance to be much smaller than the undisturbed flow, and so compared to the perturbation, we may neglect products of perturbation quantities, and the resulting equations are *linear* in the perturbations. Hence, the term "linear stability analysis". With this assumption, and substituting the above definitions into the Euler equations, we obtain the Rayleigh equation

$$(U - c)(\psi'' - k^2\psi) - U''\psi = 0. \tag{6.3}$$

This is an eigenvalue problem, where the stream function ψ is the eigenfunction, and the phase speed c is the eigenvalue, and our objective is to obtain the sign of c_i. In this flow $U = U_1$ when $z < 0$ and $U = U_2$ when $z > 0$. A perturbation of finite energy must decay far away from the interface, so we have [8]

$$\psi = A_1 e^{ikz} \quad z < 0, \tag{6.4}$$

$$\psi = A_2 e^{-ikz} \quad z > 0. \tag{6.5}$$

We also have two conditions at the interface, a dynamic condition that the normal stress must be continuous across it, giving

$$U_1 u_1 + \frac{\partial \phi_1}{\partial t} = U_2 u_2 + \frac{\partial \phi_2}{\partial t}, \tag{6.6}$$

where $\nabla\phi = u$ and we have set the surface tension to 0, and a kinematic condition

$$\frac{d\eta}{dt} = v \quad \text{giving} \quad \eta(U - c) = \psi, \tag{6.7}$$

on either side. Solving, we have

$$c = -(U_1 + U_2) \pm i\frac{U_1 - U_2}{4}, \tag{6.8}$$

giving an instability for any velocity difference, for any wavelength. In the figure we have used $U_1 = U$ and $U_2 = -U$, so the frequency of the perturbation would be zero.

6.3.2 Effect of Buoyancy: Rayleigh–Benard Flow

One of the most easily realisable experiments on hydrodynamic instabilities is that of thermal convection. Consider a mass of fluid confined between two plates a distance H apart, held at different temperatures, with the bottom plate hotter than the top. We now have an unstable stratification, with lighter fluid below heavy, and the fluid therefore has a tendency to overturn. For small temperature differences, viscosity is able to curb this tendency, and heat is transferred by conduction. As the temperature difference ΔT across a layer increases, or the diffusivities of momentum or heat (the kinematic viscosity v, and κ, respectively) decrease, the Rayleigh number $Ra \equiv g\beta\Delta T H^3/v\kappa$ increases. Here, β is the coefficient of thermal expansion. At a particular Rayleigh number of 1708 for this geometry, convection sets in and one observes patterns in the form of rolls.

6.3.3 Effect of Rotation: Taylor–Couette Flow

An adverse temperature gradient is a potentially unstable arrangement. Exploring the striking similarities between rotating and stratified flows, one can consider a fluid with unstable angular momentum "stratification". Rayleigh noted this analogy and stated that a rotating fluid will be stable to axisymmetric disturbances if its angular momentum increases monotonically outwards [3]. Consider a fluid being set into shearing motion by a pair of differentially rotating concentric cylinders (Fig. 6.4). Let us inspect two fluid parcels with different angular momenta, L_A and L_B ($L = \rho r^2 \Omega$). If one swaps A and B, then the net change in kinetic energy (centrifugal potential energy) would be

$$\Delta KE = \frac{1}{2}\left[\left\{\frac{L_B^2}{r_A^2} + \frac{L_A^2}{r_B^2}\right\} - \left\{\frac{L_A^2}{r_A^2} + \frac{L_B^2}{r_B^2}\right\}\right] = \frac{1}{2}\left(L_B^2 - L_A^2\right)\left(\frac{1}{r_A^2} - \frac{1}{r_B^2}\right).$$

$$(6.9)$$

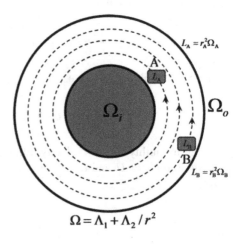

Fig. 6.4 A Taylor–Couette flow with unstable angular momentum "stratification". The constants Λ_1 and Λ_2 are determined by the inner and outer cylinder rotation rates – Ω_i and Ω_o are the radii of the cylinders

If in a certain region of the flow, L^2 decreases radially, then there would be a release of energy when the flow is perturbed, resulting in instability.

Taylor–Couette and Rayleigh–Benard are two hallmarks of the success of linear stability theory, since there is remarkable agreement between linear stability predictions and experiments. Taylor–Couette and Rayleigh–Benard also belong to a very exclusive family wherein the instability consists of oscillatory motion and takes the system away from a state of rest. Incidentally, the non-dimensional number demarcating the stable and unstable regimes in this flow, i.e., the Taylor number, is again ≈ 1700! The close analogy between Taylor–Couette and Rayleigh–Benard flows is discussed in detail in Eckhardt et al. [9], for example.

6.3.4 Effect of Surface Tension: Rayleigh–Plateau Instability

Intuitively, surface tension always tries to bring a body to a state of minimum surface energy by minimizing its surface area. For this reason, surface tension, such as viscosity, is thought to be a stabilising agent. On a plane interface, its effect can be easily envisioned. When the interface is perturbed into a wave, the surface area increases. Surface tension will then tend to suppress the undulations and bring the interface back to its original flat state. Less intuitive is the role of surface tension as a destabilising agent. Everyone has encountered a scenario wherein a smooth jet of water coming out of a tap breaks into a sequence of drops. Imagine a cylindrical jet of fluid of radius R, as shown in Fig. 6.5. The surface area is $2\pi R$ per unit length of cylinder. With a small perturbation of the interface $\eta = \eta_0 \sin(kx)$, as shown, the average radius r_1 shrinks to conserve the volume, to $r_1^2 + \eta_0^2/2 = R^2$. It is a simple exercise left to the reader to show that long wavelength perturbations would result, incredibly, in a net *decrease* in surface area, unlike on a plane interface, while short wavelength perturbations would give rise to an increase in surface area. Thus, a perturbation of long wavelength $k\eta_0 << 1$ would grow indefinitely, until a nonlinear process takes over from this instability, resulting in a break up into drops of radius r, with

$$r^3 = \frac{3\pi R^2}{2k}. \tag{6.10}$$

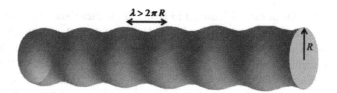

Fig. 6.5 Varicose deformations in a jet – Plateau's criterion for instability

The above physical explanation due to [9] demonstrates that a jet would spontaneously break into drops with $r > 3/2R$ to minimize surface area. Plateau realized that corrugations of a cylindrical jet beyond the perimeter of the cylinder lead to the instability, and a subsequent exhaustive study was carried forth by Rayleigh. Surface tension gradients can produce an entire class of instabilities known as Marangoni instabilities.

6.3.5 Effect of Viscosity

We discussed in the case of gravity-induced oscillations that viscosity would serve to damp out the oscillations. Indeed, it seems natural that the introduction of dissipation should lead to stabilisation. Interestingly, in shear flows, viscosity can also be the creator of instability. A pipe or channel flow without viscosity would have a constant plug-like velocity profile, which would remain stable to small perturbations. It is viscosity which creates a shear flow, i.e., a mean flow with a velocity gradient, which can then become unstable. We furnish here Rayleigh's inflexion point criterion without proof, which states that a velocity profile with an inflexion point, i.e., a location where $U'' = 0$, is a necessary condition for *inviscid* instability. The condition was made more stringent by Fjørtoft, to require a vorticity maximum as mentioned above. (The proof is available in several textbooks.) This statement seems paradoxical at first, since an inviscid flow, except a specially created one like the Kelvin–Helmholtz flow above, could typically not contain a profile of velocity, and the question of an inflexion point should not arise. The answer is simple. The adjective *inviscid* in the context of stability refers only to the study of the perturbations. The basic shear flow usually arises out of a balance between a body force and viscosity, so viscosity has a zeroth order effect on the mean flow. For example, a parabolic velocity profile is obtained in a channel since $\nu U'' = -(1/\rho)dp/dx$. In other words, viscosity, however small, cannot be neglected in the estimation of the mean flow. In the perturbation equation, on the other hand, at high Reynolds number Re, viscous effects are often *nominally* small. In particular, the Rayleigh equation, (6.3) is modified in the presence of viscosity to become the Orr–Sommerfeld equation

$$(U - c)(\psi'' - k^2\psi) - U''\psi = \frac{1}{ikRe}(\psi^{iv} - 2k^2\psi'' + k^4\psi), \qquad (6.11)$$

where the Reynolds number is defined in terms of characteristic length and velocity scales L_0 and U_0, respectively, as $Re \equiv U_0L_0/\nu$. At first sight, it appears that one may always neglect the right-hand side for sufficiently high Reynolds numbers, or sufficiently low viscosity. This is often untrue, which is why we used the word *nominally* above. A detailed explanation would take one too far away from the main purpose of this review, so we will only touch upon the problem here. One notices that the Orr–Sommerfeld is a fourth-order ordinary differential

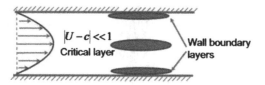

Fig. 6.6 Role of viscosity for high *Re* wall-bounded flows. The shaded regions become thinner with increasing Reynolds number, but however high the Reynolds number, viscosity may not be neglected in these regions

equation. If the right-hand side is eliminated, the order of the equation reduces by two, and all boundary conditions will not be satisfied. As illustrated in the schematic (Fig. 6.6), there are two regions where inviscid theory becomes questionable: the wall boundary layer and the critical layer – an inner boundary layer where the wave speed matches with that of flow speed at that location ($U \sim c$). Ideas explaining dissipation-induced instabilities rely on the non-trivial dynamics introduced by these viscous layers [21]. High Reynolds number instability of shear flow is thus a very good example of a singular perturbation problem. Inviscid stability results should therefore be interpreted with great care.

6.4   Absolute and Convective Instabilities

Unstable shear flows may be broadly divided into two categories: amplifiers and oscillators [14, 16–18]. An amplifier is a flow which, when you perturb by applying a force with a given frequency, will, if the flow is unstable to that frequency, feed energy from itself into the perturbation, thus making it grow. An oscillator on the other hand displays intrinsic dynamics. Apparently with no external forcing, it attains a state of self-sustained oscillations. When an amplifier is given a localised perturbation, the perturbation grows, but is simultaneously carried downstream with the flow. This means that for a short time perturbation, flow at the location where the disturbance once was, returns to its undisturbed state after some time. The speed of downstream transportation of a perturbation, which primarily contains a narrow band of frequencies is the group velocity[1] of the dominant frequency. Such an instability, where the group velocity is positive, is called convective. A typical oscillator, on the other hand, possesses one, or a small range of, frequencies whose group velocity is zero. An impulsive local perturbation would therefore grow indefinitely right at the same location (Fig. 6.7). The growth would continue while the energy of the perturbation is small, i.e., the description is linear, and non-linearities

---

[1] Group velocity denotes the speed with which a wave packet propagates, unlike phase velocity which describes rate of propagation of phase (crest or trough for a single wave). Group velocity physically denotes the rate and direction of energy propagation.

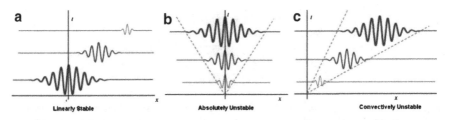

Fig. 6.7 Evolution of a disturbance wave packet. For a linearly stable flow (**a**), along all rays x/t does one observe decay. For absolutely unstable flows (**b**) the disturbance grows along the ray $x/t = 0$. (**c**) When the disturbance grows along some ray x/t but for $x/t = 0$ it decays then the flow is convectively unstable

could subsequently cause its amplitude to saturate. The frequency chosen will be that whose group velocity is zero, and the character, if any, of the initial forcing will be overwhelmed. One may well ask what the qualitative difference is between a perturbation whose group velocity is zero and one for whom it is not, since in a parallel flow, if our co-ordinate system travels downstream with the group velocity, we will observe an indefinitely growing perturbation. This would be a fair question, and the answer is that absolute instability is only interesting when there is a break in the Galilean invariance of the flow, for example due to flow non-parallelism. In a non-parallel flow, we may have one region which is absolutely unstable, and self-sustained oscillations will be created in this region. Closely related to the question of absolute instabilities is the computation of global modes in the flow system. We have so far talked about parallel flows, where the basic velocity is a function of one spatial component alone, e.g. $U = U(y)$. For the flow through a channel or pipe, well downstream of the entry region, this functional dependence is exact, but in a majority of flows, we have $U = U(x, y)$ at least. The disturbance may then be expressed in normal mode form only in z and t and not in x. We then have x being an eigendirection in addition to y, and the linear stability problem will yield two-dimensional eigenfunctions. With increasing computing power, this approach is gaining ground. It can be shown for a flow which changes a lot more weakly in x than in y that a globally unstable mode is a necessary condition for absolute instability. A more mathematical treatment is available in several of the review papers mentioned at the end of this article [4].

6.5 Non-modal Instability

As we have seen, the traditional picture of hydrodynamic instability relies on normal mode instability, with perturbations growing or decaying exponentially. Now, many common shear flows such as the flow through a pipe or a channel, either do not exhibit normal mode instability or if they do, then the Reynolds number at which normal mode instability is predicted to occur is far higher than that at which the flow is seen to become turbulent. This brings into question the whole premise of the applicability of linear stability in the transition process of such flows. The first approach was to explain away the discrepancy between theory and experiment by invoking the role of non-linearities. However, an alternative point of view has been taking shape as well. Concerted efforts over the last 30 years have shown that disturbances which decay for large times may act in unison to provide an initial transient, which grows algebraically. Mathematically, the mechanism relies on the non-normality (non-orthogonality) of the eigenmodes. We discuss below two different physical mechanisms by which a linear system can experience transient growth.

6.5.1   Orr Mechanism

More than 100 years old and probably the first description of algebraic instability in flows, the Orr mechanism can be understood purely in terms of the tilting of wave fronts by the background shear. Any passive scalar in a shear flow gets tilted and stretched along the direction of shear. Imagine a blob of ink placed in a shear flow which is initially tilted against the flow (Fig. 6.8). Advection with the flow first rotates the blob to an upright orientation and then shears it downstream, making it finer-scaled. A perturbation vorticity in two-dimensional flow would behave exactly like the ink blob and be passively advected with the flow. The first stage during which the vorticity is aligned normal to the flow requires the disturbance to extract energy from the mean flow via Reynolds stresses, and grow in the process. In the latter stage of aligning themselves along the shear, disturbances lose energy to the mean flow through the same Reynolds stresses, and decay at long times with the rate dictated by the normal modes. If one writes an equation for the perturbation

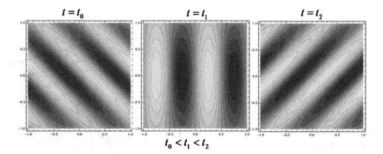

Fig. 6.8 The Orr mechanism – perturbation vorticity evolution in a Couette flow (velocity increases linearly along the vertical coordinate). An initially "up-shear" tilted perturbation is made coherent before proceeding to a "down-shear" tilt

energy density evolution for a parallel shear flow ($U(y)$), where y is the wall-normal direction and x is the streamwise direction),

$$\frac{\partial \mathscr{E}}{\partial t} = -\int_{y=a}^{y=b} U' \overline{uv} \, dy$$

$$= \int_{y=a}^{y=b} U' \overline{\frac{\partial \psi}{\partial y} \frac{\partial \psi}{\partial x}} \, dy = \int_{y=a}^{y=b} U' \overline{\left(\frac{\partial \psi / \partial x}{\partial \psi / \partial y}\right) \left(\frac{\partial \psi}{\partial y}\right)^2} \, dy$$

$$\frac{\partial \mathscr{E}}{\partial t} = -\int_{y=a}^{y=b} U' \overline{\left(\frac{\partial y}{\partial x}\right)_\psi \left(\frac{\partial \psi}{\partial y}\right)^2} \, dy \,, \qquad (6.12)$$

where $\mathscr{E}$ is the perturbation energy density and overbar denotes an average in x. Thus, energy amplification occurs ($\mathscr{E}_t > 0$) when $(\partial y / \partial x)_\psi U' < 0$ i.e. as stated before – an initial phase tilt opposite to that of mean shear.

6.5.2 Vortex Stretching–Tilting: Landahl's Lift-up Effect

In three dimensions, the perturbation vorticity is no longer a passive scalar, but one is now allowed far richer physics brought about by vortex stretching and tilting. In Fig. 6.9, the shear and vorticity of both basic flow and perturbation are depicted. The z-direction perturbation shear ($\partial_z \tilde{u}_x$, $\partial_z \tilde{u}_y$ and $\partial_z \tilde{u}_z$) acts upon the base state vorticity (into the plane, $-\bar{U}'$) to tilt and stretch it to produce $\tilde{\omega}_x$, $\tilde{\omega}_y$ and $\tilde{\omega}_z$. The mean shear also acts on $\tilde{\omega}_y$ to tilt and stretch it to produce $\tilde{\omega}_x$. This production of vorticity happens linearly in time, i.e., there is algebraic growth. An alternative explanation of this physics is that we get streamwise momentum production that is linear in time, as follows. Consider the linearized Euler equations about a base

Fig. 6.9 The base state vorticity, $-\bar{U}'$, gets stretched and tilted by perturbation shear to produce perturbation vorticity and the x and y components of perturbation vorticity feed each other through stretching–tilting by the mean shear

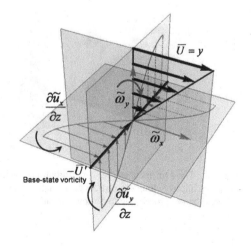

state $U(y)$ (a plane parallel flow). Additionally, assume that the disturbances have no streamwise variation.

$$\nabla^2 \tilde{p} = 0, \tag{6.13}$$

$$\frac{\partial \tilde{u}_y}{\partial t} = -\frac{\partial p}{\partial y}, \tag{6.14}$$

$$\frac{\partial \tilde{u}_x}{\partial t} = -\tilde{u}_y U'. \tag{6.15}$$

For perturbation pressure to be zero at infinity and be a harmonic function, we must have $\tilde{p} = 0$. Thus, the wall normal velocity, $\tilde{u}_y$, is independent of time and we have linear production of streamwise velocity, $\tilde{u}_x(t) = \tilde{u}_x(0) - \tilde{u}_y(0)U't$. The wall normal disturbance velocity transports ("lifts up") the mean momentum riding on its gradient to produce streamwise disturbance velocity [19].

Imagine in a shear flow that the initial disturbances are in the wall-normal velocity alone and not in the streamwise velocity. This is equivalent to having just streamwise disturbance vorticity $\tilde{\omega}_x = \partial_z \tilde{u}_y$. The vertical disturbance shear $\partial_z \tilde{u}_y$ then tilts the base state vorticity to produce wall-normal disturbance vorticity. In other words, disturbance streamwise velocity, $\tilde{\omega}_y = \partial_z \tilde{u}_x$, is produced. Thus, an initial configuration of no $\tilde{u}_x$ but containing only streamwise "rolls" ($\tilde{\omega}_x$) leads to production of $\tilde{u}_x$ or streamwise "streaks". High and low speed streaks have been observed in experiments on transition in parallel shear flows thus strengthening the notion that non-modal effects (the streak solutions are not separable in space and time and hence are not eigenmodes) are important for instability in parallel shear flows. For the three-dimensional problem both Orr and "lift-up" act together and one needs to identify the "optimal" perturbation – the initial condition which leads to maximum energy amplification [11].

6.5.3 Non-normality

We have mentioned above without explanation that for transient or algebraic growth we need the eigenmodes to be non-normal. We now discuss why non-normality is a necessary condition for secular growth of perturbations, in a situation where linear stability analysis indicates that all modes are exponentially decaying. When we examine the two parallelograms in Fig. 6.10 we see that in the second one, the sides have shrunk compared to the first, but the resultant has grown. This is because the amount of shrinking of two adjacent sides is different, and the resultant is therefore reoriented. This is reminiscent of the tilting by the Orr mechanism. One can imagine that if the shrinking of the sides of the parallelogram proceeds at an exponential rate, this growth in the resultant is only transient, and that at long time, the resultant would be zero. Now if the two adjacent sides stood for the energy of two normal modes of a given flow, and the angle between them was a measure of the degree of non-normality between the two, one can relate the growth in the

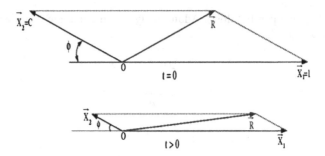

Fig. 6.10 Transient growth of the resultant of two non-orthogonal vectors

resultant to algebraic growth in the disturbance energy. If the angle was $\pi/2$, i.e., if the parallelogram were a rectangle, the resultant would decay monotonically. To show this is left to the reader as a simple exercise. Note that the eigenmodes being nonnormal is not a sufficient condition for transient growth. There are two other important ingredients: (1) the degree of nonnormality. In the parallelogram, we would need the angle between the adjacent sides to exceed $\pi - \sin^{-1}[(1-r)/(1+r)]$, where r is the ratio of their rates of exponential decay [1]. An equivalent condition may be derived for the eigenmodes in a flow system, in terms of the numerical abscissa. (2) A correct mix of initial conditions. Even when the system is sufficiently non-normal, not all combinations of initial magnitudes of the eigenmodes will result in transient growth. For a given system, we may obtain optimal perturbations which give the maximum perturbation energy at a given time. In some flows, the maximum energy can be hundreds or even thousands of times larger than the initial energy. Non-normal systems are also much more sensitive to small changes in the system matrix itself.

A vast majority of flows, i.e., almost any flow with a mean shear, is capable of displaying algebraic growth. We now recall our "success story" of linear theory, namely Rayleigh–Benard. There all the eigenmodes are orthonormal, so the question of transient growth, or of the flow leaving the laminar state to attain a new state earlier than predicted by the exponential instability, does not arise. The corresponding linear stability operators in those flows are normal, and so satisfy the condition $\mathscr{L}\mathscr{L}^A = \mathscr{L}^A\mathscr{L}$ for the orthonormality of the eigenvector basis.[2] Thus linear theory, which was earlier thought to have no role in the transition to turbulence in many

---

[2] $\mathscr{L}$ denotes some differential operator with $\mathscr{L}^A$ denoting its adjoint. They satisfy the relation, $< \mathscr{L} \leftarrow, \prec > = < \leftarrow, \mathscr{L}^{\mathscr{A}} \prec >$, where $< f, g > = \int f(x)g(x)dx$ is a typical definition for the inner product. Having already come across the Rayleigh equation $(U-c)(D^2-k^2)\psi - U''\psi = 0$, its adjoint equation can be found to be $(D^2-k^2)(U-c)\psi^\dagger - U''\psi^\dagger = 0$ with the same boundary condition. Eigenfunctions of normal operators, $\mathscr{L}\mathscr{L}^A = \mathscr{L}^A\mathscr{L}$, are orthogonal and if further $\mathscr{L} = \mathscr{L}^A$ then the operator is self-adjoint/Hermitian, which translates to all eigenvalues being real [12]. In the Orr–Sommerfeld equation, the term containing U'' contributes a term containing U' to its adjoint operator, explaining why transient growth is important in so many commonly encountered flows.

flows is now regaining many true believers! The question that remains is, if most algebraic growth is transient, and such flows are asymptotically stable, then how is the transition to turbulence affected. To complete this story, we need to discuss non-linearities. When algebraic growth has amplified the perturbation energy by a large factor, we may no longer neglect quadratic and other higher order terms in the perturbation. This means the flow becomes non-linear, follows different dynamical equations, and the asymptotic state described by linear theory is never reached. One popular description is discussed in Sect. 6.6.

6.6 Non-linearity: Self-Sustaining Processes

The question of whether non-linearity or non-normality is more relevant was posed poetically by [24] as "Nonlinear normality or nonnormal linearity?". He proposed a non-linear self sustaining process (SSP) for shear flows [25]. The already described "lift-up" effect leads to streamwise vortices ("rolls") forming alternating slow and fast "streaks". These "streaks" have an inflectional profile and suffer wake like insta- bility. This leads to a modulated streamwise flow, which further exhibits non-linear self-interaction to regenerate the streamwise "rolls". The entire process leads to self- sustained travelling waves that are constituted of counter-rotating, quasi-streamwise vortices (Fig. 6.11). These travelling wave structures have been recently observed in pipe flow experiments [15] . Most of the recent developments in this area rely on the understanding of dynamical systems, probing a state space of all velocity fields. The interested reader can refer to the review article by Eckhardt et al. [10] for details.

6.7 Viscoelastic Instabilities

Inertia promotes instabilities – this has been a driving theme for most of the in- stabilities discussed. Though viscosity induces instability for parallel shear flows, fluid with zero inertia ($Re = 0$) seems to lack an underlying mechanism to become

Fig. 6.11 Self-sustaining process in wall bounded shear flows [25]

Fig. 6.12 Extension of
polymer molecule in a
curvilinear shear flow –
(**a**) along streamlines and
(**b**) across streamlines

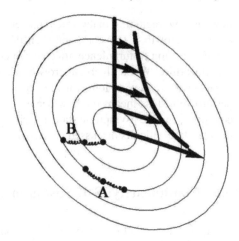

unstable. Viscoelastic flows in such a limit exhibit a zoo of instabilities. The inherent elasticity in the flow enables transition in a manner totally unknown in Newtonian flows. The well-known rod climbing effect is a manifestation of such a mechanism. In curvilinear shear flows a polymer molecule gets azimuthally stretched like iron bands on barrels producing circumferential stress known as hoop stress.

If a radial perturbation displaces the molecule from a streamline, it experiences non-uniform extension which further amplifies the hoop stress (Fig. 6.12).

Shaqfeh [23] gives a nice review on the topic. This mechanism was initially used for explaining the instability of inertialess viscoelastic Taylor–Couette flow. Groisman and Steinberg [13] went further and observed that viscoelastic flow confined between differentially rotating discs exhibited signatures of turbulence at $Re = 10^{-15}$! This was termed as "elastic turbulence" and since then has been an active area of research.

The above arguments for elastic instability rely on the base flow having curvilinear streamlines. However, many instabilities in the context of parallel shear flows could not be explained through this argument. It was realised that finite amplitude disturbances, which induce local curvature in the perturbed flow could allow for the aforementioned instability – a subcritical instability [22].

References

1. Bale R, Govindarajan R (2010) Transient Growth and why we should care about it, Resonance, 15:441–457
2. Batchelor GK (1967) Introduction to fluid dynamics. Cambridge University Press, Cambridge
3. Chandrasekhar S (1981) Hydrodynamic and hydromagnetic stability. Dover Publications, New York
4. Chomaz JM (2005) Global instabilities in spatially developing flows: Non-normality and non-linearity. Annu Rev Fluid Mech 37:357–392
5. Craik ADD (2004) The origins of water wave theory. Annu Rev Fluid Mech 36:1–28

6. Craik ADD (2005) George Gabriel Stokes on water wave theory. Annu Rev Fluid Mech 37: 23–42
7. de Gennes PG, Brochard F, Quere D (2004) Capillarity and wetting phenomena: Drops, bubbles, pearls, waves. Springer, New York
8. Drazin PG (2002) Introduction to hydrodynamic stability. Cambridge University Press, Cambridge
9. Eckhardt B, Grossman S, Lohse D (2007) Torque scaling in turbulent Taylor–Couette flow between independently rotating cylinders. J Fluid Mech 581:221–250
10. Eckhardt B, Schneider TM, Hof B et al (2007) Turbulence transition in pipe flow. Annu Rev Fluid Mech 39:447–468
11. Farrell BF, Ioannou PJ (1993) Optimal excitation of three-dimensional perturbations in viscous constant shear flow. Phys Fluids A 5:1390–1400
12. Friedman B (1990) Principles and techniques of applied mathematics. Dover, New York
13. Groisman A, Steinberg V (2000) Elastic turbulence in a polymer solution flow. Nature 405:53–55
14. Ho CM, Huerre P (1984) Perturbed free shear layers. Annu Rev Fluid Mech 16:365–422
15. Hof B, van Doorne CWH, Westerweel J et al (2004) Experimental observation of nonlinear traveling waves in turbulent pipe flow. Science 305:1594–1598
16. Huerre P, Monkewitz PA (1990) Local and global instabilities in spatially developing flows. Annu Rev Fluid Mech 22:473–537
17. Huerre P, Rossi M (1998) Hydrodynamic instabilities in open flows. In: Godreche C, Manneville P (eds) Hydrodynamics and nonlinear instabilities 81–294, Cambridge University Press, Cambridge
18. Huerre P (2000) Open shear flow instabilities. In: Batchelor GK, Moffatt HK, Worster MG (eds) Perspectives in fluid dynamics: A collective introduction to current research, Cambridge University Press, London
19. Landahl MT (1980) A note on an algebraic instability of inviscid parallel shear flows. J Fluid Mech 98:243–251
20. Lighthill MJ (1978) Waves in fluids. Cambridge University Press, London
21. Lindzen RS (1988) Instability of plane parallel shear flow (toward a mechanistic picture of how it works). PAGEOPH 126:103–121
22. Morozov AN, van Saarloos W (2007) An introductory essay on subcritical instabilities and the transition to turbulence in visco-elastic parallel shear flows. Phys Rep 447:112–143
23. Shaqfeh ESG (1996) Purely elastic instabilities in viscometric flows. Annu Rev Fluid Mech 28:129–185
24. Waleffe F(1995) Transition in shear flows. Nonlinear normality versus non-normal linearity. Phys Fluids 7:3060–3066
25. Waleffe F (1997) On a self-sustaining process in shear flows. Phys Fluids 9:883–900

Further Reading

1. Bayly BJ, Orszag SA, Herbert T (1988) Instability mechanisms in shear-flow transition. Annu Rev Fluid Mech 20:359–391
2. Craik ADD (1985) Wave interaction and fluid flows. Cambridge University Press, Cambridge
3. Drazin PG, Reid WH (1982) Hydrodynamic stability. Cambridge University Press, Cambridge
4. Grossman S (2000) The onset of shear flow turbulence. Rev Mod Phys 72: 603–618
5. Monin AS, Yaglom AM (1997) Statistical fluid mechanics, The mechanics of turbulence. New English edn. CTR Monographs, NASA Ames Stanford University 1:2–4
6. Schmid PJ, Henningson DS (2001) Stability and transition in shear flows. 1st edn. Springer-Verlag, New York
7. Schmid PJ (2007) Nonmodal stability theory. Annu Rev Fluid Mech 39:129–162
8. Stuart JT (1958) On the nonlinear mechanics of hydrodynamic stability. J Fluid Mech 4:1–21
9. Stuart JT (1971) Nonlinear stability theory. Annu Rev Fluid Mech 3:347–370

Part III
Applications

Chapter 7
Statics and Dynamics of Dilute Polymer Solutions

Arti Dua

Abstract Polymers show universal behaviour at long length and time scales. In this chapter, the statics and dynamics of dilute polymer solutions are presented in terms of theoretical models that form the basis of polymer physics. The size of an ideal polymer is calculated from the freely jointed chain model, the freely rotating chain model and the Gaussian equivalent chain model. The Edwards Hamiltonian for a continuous ideal chain is obtained as a continuum limit of the Gaussian equivalent chain. The size of a real chain in good and poor solvents is estimated from the free energy that includes entropic and enthalpic contributions. The temperature dependence of chain size is discussed using the concept of thermal blobs. The force–extension relations for ideal and real polymers are illustrated using tension blobs. A brief introduction to stiff chains is presented in terms of the worm-like chain model. Starting from the Langevin equation for a Brownian particle, the salient features of the Rouse and Zimm models are presented. The behaviour of polymers under extensional, rotational, simple shear and linear-mixed flows is discussed using the finitely extensible Rouse chain.

7.1 Introduction

Polymers are large molecules composed of many smaller repeating units (monomers) bonded together. Solvent-induced thermal fluctuations bring about the internal rotation of bonds generating large numbers of conformational states [2, 13, 14, 37]. On time scales much larger than the rotational times, experimental measurements typically capture the average effects of the rapid fluctuations in the conformations. Even instantaneous measurements involve averages over large ensembles of identical molecules. Given that the system has a large number of

A. Dua (✉)

Department of Chemistry, Indian Institute of Technology Madras, Chennai 600036, India
e-mail: arti@iitm.sc.in

J.M. Krishnan et al. (eds.), *Rheology of Complex Fluids*,
DOI 10.1007/978-1-4419-6494-6_7, © Springer Science+Business Media, LLC 2010

degrees of freedom, there is no general methodology, either classical or quantum mechanical, that can treat the system exactly. Apart from the enormously large number of conformations, each polymeric system is characterized by its own unique chemistry at the microscopic level because of the different chemical structures of the monomers. Nevertheless, at one level, polymeric systems can be quite simple, as the following suggests: the macroscopic properties of a collection of small molecules depend strongly on the intrinsic chemical details of the molecules, but when the same set of chemical species is bonded to form long molecules, they exhibit universal behaviour. In other words, at length scales much larger than the monomer size, the properties are insensitive to the structure of the individual molecules of the chain, but depend on the universal chain-like nature of the polymer. Thus, long wavelength properties can be described using general statistical mechanical methods that depend on a few system dependent phenomenological parameters [2, 5–7, 13, 14, 16, 17, 32, 37, 38]. In what follows, we describe theoretical methods that form the basis of polymer physics. Minimal models describing long lengthscale chain-like properties of polymers that depend on a few phenomenological parameters are presented in the next section.

7.2 Ideal Polymers

An equilibrium property that governs the behaviour of a polymer solution is the chain size and its dependence on temperature. The understanding of chain structure is the key to unraveling the mechanical, thermodynamical and dynamical properties of polymer solutions. The size of a polymer depends on the quality of solvent in which it is dissolved. In a real polymer, the relative strengths of the monomer–monomer and monomer–solvent interactions determine if the chain is swollen or collapsed. At high temperatures, the effective monomer–monomer interactions are repulsive and polymers are found in a swollen state. At low temperatures, polymers form globules due to dominance of attractive interactions. At a special temperature, called the θ-temperature, the attractive and repulsive parts of the effective monomer–monomer interactions cancel each other resulting in nearly ideal chain conformations. It is worth mentioning that concentrated polymer solutions and linear polymer melts show near ideal chain behaviour since surrounding polymers screen the interactions between monomers. The conformational behaviour of an ideal polymer forms the basis of most models in polymer physics [2, 5–7, 13, 14, 16, 17, 32, 37, 38]. In what follows, we describe several different models that estimate the size of an ideal polymer.

The end-to-end vector, $\mathbf{R}$, of a flexible polymer consisting of n bond vectors is given by

$$\mathbf{R} = \sum_{i=1}^{n} \mathbf{r}_i, \tag{7.1}$$

Fig. 7.1 Freely jointed chain

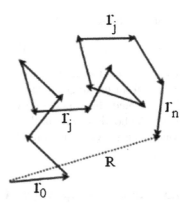

where $\mathbf{r}_i$ is the ith bond vector of magnitude $|\mathbf{r}_i| = l$. The mean square end-to-end distance of this chain is given by

$$\langle \mathbf{R}^2 \rangle = \sum_{i=1}^{n} \sum_{j=1}^{n} \langle \mathbf{r}_i \cdot \mathbf{r}_j \rangle = l^2 \sum_{i=1}^{n} \sum_{j=1}^{n} \langle \cos \theta_{ij} \rangle, \tag{7.2}$$

where θ_{ij} represents the angle between the bond vectors $\mathbf{r}_i$ and $\mathbf{r}_j$ and $\langle \cdots \rangle$ corresponds to the ensemble average. Figure 7.1 represents a freely jointed chain. In a freely jointed chain model, all bond vectors are completely uncorrelated with each other, that is, $\langle \cos \theta_{ij} \rangle = 0$ for $i \neq j$. This leads to a very simple expression for the mean square distance, $\langle \mathbf{R}^2 \rangle = nl^2$. In general, restricted bond-angle rotations and steric hindrance lead to local correlations between bond vectors along the chain backbone. These local correlations are accounted for by defining Flory's characteristic ratio,

$$C_n = \frac{1}{n} \sum_{i=1}^{n} \sum_{j=1}^{n} \langle \cos \theta_{ij} \rangle. \tag{7.3}$$

In terms of Flory's characteristic ratio, the mean square distance is given by $\langle \mathbf{R}^2 \rangle = C_n n l^2$. Flory's characteristic ratio, C_n, increases with the increase in the number of monomers. For $n \gg 1$, it saturates to a finite value C_∞ yielding $\langle \mathbf{R}^2 \rangle \approx C_\infty n l^2$. For many flexible polymers, C_∞ is typically 7–9.

A simple estimate of C_∞ is provided by the freely rotating chain model. The freely rotating model assumes all torsional angles to be equally probable. It accounts for the local correlation by estimating $\langle \mathbf{r}_i \cdot \mathbf{r}_j \rangle = l^2 (\cos \theta)^{|i-j|}$. In terms of the exponential function, $(\cos \theta)^{|i-j|} = \exp(-|i - j|l/l_p)$, where $l_p = -l/\ln(\cos \theta)$ defines the length scale at which the correlations between the bond vectors decay. Thus, the mean-square end-to-end distance of a freely rotating chain is given by

$$\langle \mathbf{R}^2 \rangle = l^2 \sum_{i=1}^{n} \sum_{j=1}^{n} \exp(-|i - j|l/l_p), \tag{7.4}$$

which simplifies to yield

$$\langle R^2 \rangle = nl^2 \frac{1 + \cos\theta}{1 - \cos\theta}, \tag{7.5}$$

where $C_\infty = (1 + \cos\theta)/(1 - \cos\theta)$. For flexible polymers, the value of C_∞ calculated from the freely rotating chain model is less than expected. However, the freely jointed chain model provides a rough estimate of the local correlations between the bond vectors. In the next section, a unified description of ideal flexible polymers is provided in terms of the Gaussian equivalent chain.

7.3 Gaussian Equivalent Chains

As a rough approximation, the conformation of a flexible chain can be viewed as the trajectory of a random walker that makes n independent steps of fixed step length l to cover a certain distance R [3, 29]. What is the probability for R? Since the steps are completely uncorrelated with respect to each other, the overall probability of a given conformation is the product of the probabilities of the individual steps, that is,

$$P = \prod_0^n p(\mathbf{r}_i), \tag{7.6}$$

where $\mathbf{r}_i$ is the bond vector of the ith step. Since the magnitude of the bond (step) length is fixed, the simplest way to define the bond probability is

$$p(\mathbf{r}_i) = \frac{1}{4\pi l^2} \delta(|\mathbf{r}_i| - l). \tag{7.7}$$

The bond probability defined above accounts for the constraint of a constant bond length, but it leads to a rather complicated distribution function for $\mathbf{R}$, the vectorial distance from one end of the chain to the other. At the same time, since $\mathbf{R}$ is the sum of a large number of random variables (bond vectors), the probability function can be described by a Gaussian function according to the central limit theorem. For large enough n and distance much less than nl, the distribution is given by

$$P(\mathbf{R}, n) = \left(\frac{3}{2\pi n l^2}\right)^{3/2} \exp\left(-\frac{3\mathbf{R}^2}{2n l^2}\right). \tag{7.8}$$

The above distribution can be exactly produced by a bond probability that is given, not by (7.1), but by

$$p(\mathbf{r}_i) = \left(\frac{3}{2\pi l^2}\right)^{3/2} \exp\left(-\frac{3\mathbf{r}_i^2}{2 l^2}\right). \tag{7.9}$$

A chain described by the above bond distribution function is referred to as a Gaussian chain. Since the Gaussian chain is made up of several Gaussian links, each effective step of the random walk is itself a random walk. The Gaussian chain is, therefore, a self-similar object, or fractal [27]. Its structure looks the same on any length scale. If we group n_0 monomers together to form a single unit, the polymer chain is again a random walk of larger step length. To be specific, there are $N = n/n_0$ effective steps of step length $b = n_0^{1/2} l$, with $\langle \mathbf{R}^2 \rangle = N b^2 = n l^2$. In other words, averaging over certain degrees of freedom (coarse-graining) does not change the general scaling behaviour of the long wavelength properties of the chain. In particular, the distribution function depends on the ratio $\mathbf{R}^2 / \langle \mathbf{R}^2 \rangle$, which is scale invariant. The step length defined above is a coarse-grained length, which is the average over a large number of successive bond vectors. The chain thus obtained is a coarse-grained description of a real chain, and is equivalent to it at this level of description. The chain can, therefore, be called an equivalent Gaussian chain. The notion of equivalent chains was first introduced by Werner Kuhn [16, 17]. The coarse-grained length b is hence commonly referred to as the Kuhn length. In terms of the bond length l and the number of monomers n, the Kuhn length and the effective number of monomers are given by $b = C_\infty l / \cos(\theta/2)$ and $N = n \cos^2(\theta/2)/C_\infty$. As long as the Kuhn length is much smaller than the contour length, an equivalent chain describes a flexible polymer. An equivalent chain is a good approximation of a real chain for long wavelength properties. Indeed, most theoretical approaches model flexible chains in terms of the equivalent chain description.

The equivalent Gaussian chain has the unphysical feature that it can extend more than its contour length. In fact, the chain has an infinite stored length. This feature is the reflection of the underlying Gaussian nature of the distribution function, which has a long length tail, and hence a finite probability for extension beyond its contour length. The infinite extensibility is often a serious limitation when describing chain extension under various kinds of force fields. Some of these issues are discussed in Sect. 7.9, where a finitely extensible Gaussian model is presented for chain extension under shear flow.

7.4 Continuous Chains

The equivalent Gaussian chain can be subjected to a further level of coarse-graining by letting $b \to 0$, $N \to \infty$ and $R_{\max} = Nb$ constant. This leads to a representation of the polymer as a continuous path with fixed ends. The probability that such a path begins at $\mathbf{R}_0$ and ends at $\mathbf{R}$ is given by a path integral representation:

$$G(\mathbf{R}, \mathbf{R}_0; N) = \int_{\mathbf{r}_0 = \mathbf{R}_0}^{\mathbf{r}_N = \mathbf{R}} D[\mathbf{r}_n] \exp\left[-\frac{3}{2b^2} \int_0^N dn |\dot{\mathbf{r}}_n|^2 \right], \qquad (7.10)$$

where $\mathbf{r}_n$ is the position vector of the $n^t h$ monomer. N is the effective number of monomers in the polymer. The integral over $D[\mathbf{r}_n]$ is a continuous representation of the discrete sum over all possible chain conformations, and it contains factors that ensure that $G(\mathbf{R}, \mathbf{R}_0; N)$ is normalized. The advantages of going over to a continuum chain description lie in the fact that path integrals can be evaluated using established techniques in quantum and statistical mechanics [12, 34]. Moreover, the end-vector distribution function defined above satisfies a diffusion equation that is formally equivalent to the Schrodinger equation in quantum mechanics. The formal analogy is useful in treating many of the problems concerning a polymer in an external field.

Equation (7.10) can be recognized as the Weiner probability distribution. Although the chain conformation is considered to be a continuous curve, the continuous limit is mathematically subtle since the curve is not differentiable in the classical sense [20, 28]. In other words, the curve does not have a well-defined tangent at any point. This happens to be an inherent property of a self-similar object. However, (7.10) leads to a physically meaningful probability distribution that accounts for connectivity of the chain, which is completely entropic in origin. The "Hamiltonian" for the flexible polymer is of the form:

$$\beta H = \frac{3}{2b^2} \int_0^N dn \, |\dot{\mathbf{r}}_n|^2, \qquad (7.11)$$

where $\beta = 1/k_B T$. Since the above Hamiltonian only accounts for the chain connectivity, it represents an ideal chain Hamiltonian. If the polymer interacts with the external field, the end-to-end vector distribution function of (7.10) is modified to

$$G(\mathbf{R}, \mathbf{R}_0; N) = \int_{\mathbf{r}_0=\mathbf{R}_0}^{\mathbf{r}_N=\mathbf{R}} D[\mathbf{r}_n] \exp \left[- \int_0^N \Delta n \left(\frac{3}{2b^2}|\dot{\mathbf{r}}_n|^2 + V[\mathbf{r}_n] \right) \right], \quad (7.12)$$

where $V[\mathbf{r}_n]$ may be a solvent field, a gravitational field or a field due to intramolecular excluded volume interactions between different parts of the same chain. Non-local excluded volume and three-body interactions account for the quality of solvents in which polymers are dissolved. These interactions are often described by two-body and three-body pseudo-potentials [6, 7, 16, 17] given by

$$\beta H = \frac{3}{2b^2} \int_0^N \Delta n \left(\frac{\partial \dot{\mathbf{r}}_n}{\partial n} \right)^2 - \frac{v}{2} \int_0^N \Delta n \int_0^N \Delta m \delta[\mathbf{r}_n - \mathbf{r}_m]$$
$$+ \frac{w}{6} \int_0^N \Delta n \int_0^N \Delta m \int_0^N \Delta l \delta[\mathbf{r}_n - \mathbf{r}_m] \delta[\mathbf{r}_m - \mathbf{r}_l], \qquad (7.13)$$

where the first term describes the entropic elasticity of the chain; the second and the third terms represent the two-body repulsive excluded volume and three-body repulsive interactions, respectively; v and w represent the strengths of the two-body and three-body interactions, respectively. The size of a flexible polymer is characterized by its mean square radius of gyration, $\langle \mathbf{R}_g^2 \rangle$. For a linear and flexible polymer, $\langle \mathbf{R}_g^2 \rangle = \langle \mathbf{R}^2 \rangle / 6$, where $\langle \mathbf{R}^2 \rangle$ is the mean square end-to-end distance. The mean square end-to-end distance can be evaluated from the following expression:

$$\langle \mathbf{R}^2 \rangle = \frac{\int_{-\infty}^{\infty} d\mathbf{R} \, \mathbf{R}^2 \exp(-\beta H)}{\int_{-\infty}^{\infty} d\mathbf{R} \, \exp(-\beta H)}. \tag{7.14}$$

An ideal chain Hamiltonian is given by (7.11). The evaluation of the chain size involves a simple Gaussian integration, which yields $\langle \mathbf{R}^2 \rangle = Nb^2$. A real chain Hamiltonian, described by (7.13), involves both entropic and enthalpic contributions. The evaluation of the chain size is, therefore, a difficult task. In the next section, we provide a brief overview of the Flory theory, which provides an estimate of the chain size in the presence of enthalpic interactions [13, 14].

7.5 Temperature Dependence of Polymer Size

The size of a polymer depends on the effective interactions between monomers, which are characterized by excluded volume (second virial coefficient) and three-body interaction parameters. The excluded volume parameter v estimates the effective interaction between the monomers, $v = - \int d^3 r f$, where the Mayer f-function is $f = \exp(-\beta H_c) - 1$, and H_c represents the interaction between different monomers. A similar treatment for three-body potentials yields an expression for w. Typical average values of these parameters are $v \sim b^3 (1 - \theta / T)$ and $w \sim b^6$, where θ is the temperature at which the chain is ideal [5–7, 16, 17, 32]. Away from the θ temperature, the sign of the excluded volume parameter v determines whether the effective two-body interactions are repulsive or attractive. For $T > \theta$, the effective interactions between monomers are repulsive. At temperatures higher than the θ temperature, therefore, monomers like the solvent in which they are dissolved and the solvent is termed a good solvent. In the opposite limit of $T < \theta$, the effective two-body interaction is attractive and the solvent is termed a poor solvent.

The size of a real polymer was first estimated by Flory by identifying the entropic and energetic contributions to the total free energy $F(\mathbf{R}, N)$,

$$F(\mathbf{R}, N) = F_{\text{ent}}(\mathbf{R}, N) + F_{\text{int}}(\mathbf{R}, N), \tag{7.15}$$

where N is the effective number of monomers and $\mathbf{R}$ is the end-to-end vector.

The entropic contribution to the free energy can be calculated from the probability distribution for end-to-end vector of an ideal polymer which follows Gaussian statistics,

$$P(\mathbf{R}, N) = \left(\frac{3}{2\pi N b^2} \right)^{3/2} \exp(-3\mathbf{R}^2 / 2Nb^2), \tag{7.16}$$

resulting in the following entropic contributions to the free energy:

$$S(\mathbf{R}, N) = k_B \ln P(\mathbf{R}, N) \tag{7.17}$$

$$= S(0, N) - \frac{3k_B T \mathbf{R}^2}{2Nb^2}. \tag{7.18}$$

The free energy of a polymer is given by $F(R, N) = U(R, N) - TS(R, N)$. For an ideal polymer there are no energetic contributions, and the entropic free energy is given by $F_{ent}(R, N) = \frac{3k_B T R^2}{2Nb^2}$. For a real polymer, the energetic contribution to the free energy per unit volume can be calculated by carrying out a virial expansion in powers of the monomer number density c_n. The coefficient of the c_n^2 term is proportional to the excluded volume v and the coefficient of the c_n^3 is related to the three-body interaction coefficient $w \approx b^6$,

$$\frac{F_{int}}{V} = \frac{k_B T}{2} \left(vc_n^2 + wc_n^3 + \cdots \right) \approx \frac{k_B T}{2} \left(v\frac{N^2}{R^6} + w\frac{N^3}{R^9} + \cdots \right). \tag{7.19}$$

The total free energy, with contributions from entropic and enthalpic terms, is given by

$$\frac{F(R, N)}{k_B T} \approx \frac{R^2}{Nb^2} + v\frac{N^2}{R^3} + w\frac{N^3}{R^6}. \tag{7.20}$$

From the above equation, the size of a real polymer in good and poor solvents can be evaluated.

In a good solvent, the repulsive nature of the effective monomer–monomer interactions swells the chain. Since under these conditions, the three-body interaction term is not important, the size of the polymer can be evaluated by just retaining the entropic and the repulsive excluded volume interaction terms,

$$\frac{F(R, N)}{k_B T} \approx \frac{R^2}{Nb^2} + v\frac{N^2}{R^3}. \tag{7.21}$$

By minimizing the above free energy with respect to R, i.e. $\frac{\partial F(R,N)}{\partial R} = 0$, the size of the polymer in a good solvent can be estimated. The equilibrium size is given by $R \approx N^{3/5} v^{1/5} b^{3/5}$. The latter is a universal expression for the chain size in a good solvent. In contrast to the exponent of 0.6, the more sophisticated renormalization group theories estimate this exponent as 0.588 [16, 17]. The Flory theory, in spite of its simplicity, predicts the right scaling for the chain size.

In a poor solvent, the effective attractive monomer–monomer interactions lead to a collapsed globular structure. Since the resulting structure is highly compact, the entropic contributions are not important and the size of the polymer can be evaluated by just retaining the three-body repulsive and the two-body attractive excluded volume interaction terms,

$$\frac{F(R, N)}{k_B T} \approx -|v|\frac{N^2}{R^3} + w\frac{N^3}{R^6}. \tag{7.22}$$

By minimizing the above free energy with respect to R, the equilibrium size in a poor solvent can be estimated as $R \approx (N/|v|)^{1/3}b$. Therefore, a globule-to-coil conformational transition can be brought about by changing the sign of the excluded volume interaction term, which amounts to changing the solvent quality or temperature.

To understand the temperature dependence of a real polymer, it is customary in polymer physics to introduce the concept of blobs [5]. Since polymers are large molecules, there exists a separation of length scales within the polymer structure. Irrespective of the nature of the interaction energy, the size of a blob corresponds to the length scale at which the interaction energy is of the order of the thermal energy, $k_B T$. On length scales smaller than the blob size, the chain follows the unperturbed statistics. On length scales larger than the blob size, the chain statistics is determined by the interaction energy.

The free energy of a real polymer has entropic and enthalpic contributions. The length scale at which the free energy is of the order of $k_B T$ determines the size of a thermal blob, ξ_T. The entropic contribution to the free energy is $\beta F_{\text{ent}} = R^2/Nb^2$. Since length scales below the blob size are dominated by entropy, we have $\xi_T \sim n_T^{1/2}b$ where n_T is the number of monomers in a thermal blob. To determine n_T, we set $\beta F_{\text{int}} \sim 1$, which yields $n_T \sim b^6/v^2$. For length scales larger than ξ_T, the thermal blobs are self-avoiding in a good solvent and space-filling in a poor solvent. In terms of the thermal blob, therefore, the sizes in good and poor solvents are given by $R \sim (N/n_T)^{3/5}\xi_T \sim N^{3/5}v^{1/5}b^{3/5}$ and $R \sim (N/n_T)^{1/3}\xi_T \sim (N/|v|)^{1/3}b$. The latter expressions are the same as obtained from the Flory theory.

7.6 Polymers Under Tension

The free energy of an ideal polymer has only entropic contribution given by $F(\mathbf{R}, N) = 3k_B T \mathbf{R}^2/2Nb^2$. To hold the chain at a fixed end-to-end vector $\mathbf{R}$, one needs to apply equal and opposite force at the chain ends,

$$\mathbf{f} = \frac{\partial F(\mathbf{R}, N)}{\partial \mathbf{R}} = \frac{3k_B T}{Nb^2}\mathbf{R}. \tag{7.23}$$

The force–extension relation is linear and follows Hooke's law for small deformations. This is the result of the Gaussian approximation for the end-to-end vector distribution. The above relation suggests that the spring constant is $k = 3k_B T/Nb^2$. The dependence of the spring constant on temperature is a reflection of the entropic elasticity.

The linear relation for ideal chains can also be obtained from the concept of blobs [5, 32]. The length scale at which the free energy due to the applied force is of the order of $k_B T$ defines the tension blob, $\xi_f \sim k_B T/f$. For length scales shorter than the tension blob, the chain statistics is unperturbed and is governed by entropy, that is, $\xi_f \sim n_f^{1/2}b$, where n_f is the number of monomers in the tension blob. For length scales larger than the tension blob, the chain statistics is determined by the applied force yielding a stretched state $R \sim (N/n_f)\xi_f \sim Nb^2 f/k_B T$. The latter suggests that for an ideal chain $f \sim (k_B T/Nb^2)R$. This recovers the linear relation between force and extension in agreement with (7.23).

For real chains at high temperatures, the unperturbed statistics is that of a self-avoiding walk. For length scales shorter than the tension blob, therefore, $\xi_f \sim n_f^{3/5}b$. At large length scales, the chain is in a stretched state given by $R \sim (N/n_f)\xi_f \sim Nb(fb/k_B T)^{2/3}$ implying $f = k_B T/b(R/Nb)^{3/2}$. In contrast to ideal chains, the force–extension relation for real chains is non-linear. The scaling arguments discussed in this section are only valid for small deformations. For large deformations, the force–extension relation can be derived from the worm-like chain (WLC) model. In the next section, the chain statistics of stiff polymers is presented in terms of the wormlike chain model.

7.7 Stiff Chains

The discussion so far has been limited to a class of polymers that are completely flexible. The flexibility in such molecules arises because of the unhindered rotation of bond vectors through a fixed bond angle. In stiff polymers, like double-stranded DNA, the flexibility is due to the fluctuations of the contour length about a straight line. The WLC is a continuous chain description of stiff polymers. The WLC model is a special case of the freely rotating model for small bond angles. The concept of the wormlike chain was first introduced by Kratky and Porod. The wormlike chain is also referred to as the Kratky-Porod Chain. The mean-square distance of the wormlike chain can be calculated from (7.4). Since the WLC is a continuous description, the summations in (7.4) can be expressed as integrals,

$$\langle \mathbf{R}^2 \rangle = \int_0^L ds' \int_0^L ds \, \exp(-|s - s'|/l_p), \tag{7.24}$$

where $L = nl = Nb$, $l_p = 2l/\theta^2$ and $b = 2l_p$ for $\theta \ll 1$; the resultant expression is given by

$$\langle \mathbf{R}^2 \rangle = 2Ll_p - 2l_p^2(1 - \exp(-L/l_p)). \tag{7.25}$$

The persistence length, l_p, is a measure of the distance over which the bond vectors are correlated, and hence determines the effective stiffness of the chain. That (7.25) is a valid description in both stiff and flexible limits can be seen as follows: when $Ll_p \ll 1$, $\langle \mathbf{R}^2 \rangle = L^2$ (the rod limit); when $Ll_p \gg 1$, $\langle \mathbf{R}^2 \rangle = 2Ll_p = Nb^2$ (the flexible limit).

For a chain with a constant contour length, there exists a unit tangent vector given by $\mathbf{u}(s) = \partial \mathbf{r}(s)/\partial s$, where $\mathbf{r}(s)$ is the position of a point at arc length s from one end of the chain. For a stiff chain, the orientation of the unit vector $\mathbf{u}(s)$ changes slowly with respect to the arc length variable s. In general, for chains with variable stiffness, the bending energy is proportional to the change in the orientation of the unit vector with respect to the arc length variable s, $\partial \mathbf{u}(s)/\partial s$. Thus, the lowest order scalar function of $\partial \mathbf{u}(s)/\partial s$ that can contribute to the bending energy must be quadratic in this variable. Since $\mathbf{u}(s) \cdot \partial \mathbf{u}(s)/\partial s = 0$, the only quadratic form is

$$H = \frac{\varepsilon}{2} \int_0^L ds \left(\frac{\partial \mathbf{u}(s)}{\partial s} \right)^2, \tag{7.26}$$

where ε is the elastic energy for bending, which is equal to $k_B T l_p$. The connectivity term in the above Hamiltonian is absent because of the constraint $|\mathbf{u}(s)| = 1$, which follows from simple differential geometry. Since the orientation of the unit tangent vector varies slowly along the stiff chain backbone, the unit tangent vectors are correlated over longer distances compared to a flexible chain. In general, the correlations between the unit tangent vectors at two different points along the chain backbone decay exponentially, and are of the form

$$\langle \mathbf{u}(s) \cdot \mathbf{u}(s') \rangle = \exp(-L/l_p). \tag{7.27}$$

Saito, Takahashi and Yunoki have estimated the tangent vector distribution function in terms of an infinite series, with the above constraint on the unit tangent vector [33]. Their distribution provides a correct estimate for the second and fourth moments of the end-to-end distance, R. However, the probability distribution for the end-to-end distance cannot be calculated within their formalism, even in terms of an infinite series. To simplify the analysis, Harris and Hearst relaxed the local constraint of constant unit tangent vectors to a global constraint of constant contour length by introducing a Lagrange multiplier, and treating it as a free parameter [18]. Although their approach reproduces the mean-square end-to-end distance by fixing the free parameter, it fails to yield other second moment quantities except in the rod and flexible limits. In a different approach, Freed relaxed the local constraint $|\mathbf{u}(s)| = 1$ to statistical constraint $\langle \mathbf{u}(s)^2 \rangle = 1$, producing a rather simple model in which the chain Hamiltonian has a connectivity term along with a bending energy term [15]. The coefficient in front of the bending energy term is kept as a free parameter and is fixed by requiring that $\langle \mathbf{u}(s)^2 \rangle = 1$. For this model, the probability distribution function that the chain ends are at 0 and $\mathbf{R}$ with tangent vectors $\mathbf{U}'$ and $\mathbf{U}$ is of the form

$$G(\mathbf{R}, 0; \mathbf{U}, \mathbf{U}'; N, 0) = \int_{\mathbf{u}(0)=\mathbf{U}'}^{\mathbf{u}(L)=\mathbf{U}} D[\mathbf{u}(s)] \delta \left[\mathbf{R} - \int_0^L ds\, \mathbf{u}(s) \right]$$

$$\times \exp \left[-\left(\frac{3}{4l_p} \int_0^L ds\, |\mathbf{u}(s)|^2 + \frac{l_p}{2} \int_0^L ds\, |\dot{\mathbf{u}}(s)|^2 \right) \right]. \tag{7.28}$$

However, the distribution function estimates $\langle \mathbf{R}^2 \rangle = L l_p$, which is just the flexible chain result. There is no dependence on the bending energy ε. This spurious behaviour is because of the inhomogeneous nature of the chain defined by (7.28). The inhomogeneity becomes more explicit when the mean square fluctuations, $\langle \mathbf{u}(s)^2 \rangle$, of the tangent vectors are calculated at different points along the chain backbone. It turns out that there is as much as a factor of 2 increase in $\langle \mathbf{u}(s)^2 \rangle$ for s at the ends of the chain as compared to s at the middle [24]. Therefore, the excess fluctuations at the chain ends need to be suppressed by incorporating an energy penalty for fluctuations at the ends. A Hamiltonian that does that is given by

$$\beta H = \frac{3}{4l_p} \int_0^L ds |\mathbf{u}(s)|^2 + \frac{3l_p}{4} \int_0^L ds |\dot{\mathbf{u}}(s)|^2 + \frac{3}{4}(|\mathbf{u}_0|^2 + |\mathbf{u}_N|^2). \quad (7.29)$$

The probability distribution for the end-to-end distance derived from this Hamiltonian provides correct predictions for various second moments. In particular, it reproduces the predictions of the Kratky–Porod model while remaining analytically tractable.

7.8 Rouse and Zimm Models

The diffusion of a large molecule in a solvent of small molecules generally follows Brownian motion. Since a Brownian particle is much heavier than the surrounding fluid particles, there exists a wide separation of time scales in their respective microscopic dynamics. The Brownian particle changes its velocity on a much slower time scale; the fluid particles, on the other hand, change their velocities on a much faster time scale. As a result, the Brownian particle suffers a large number of collisions with the fluid particles in its path through the fluid. It is, therefore, possible to identify the velocity of the Brownian particle as a slow mode of the system, and the velocities of the fluid particles as fast modes of the system. On time scales much longer than the frequency and duration of individual collisions, the dynamics of the Brownian particle can be expressed in terms of an equation in which the details of the collision process are contained in the statistical properties of a stochastic variable (the fast modes). Such an equation was suggested by Langevin [25], and is given by

$$m\ddot{\mathbf{r}}(t) = -\zeta \dot{\mathbf{r}}(t) + \mathbf{f}(t). \quad (7.30)$$

The first term in this equation represents the inertial force. The second term is the systematic force with a mean $-\zeta \mathbf{v}$, where ζ is the friction coefficient. The systematic force is the average frictional force, and is proportional to the velocity of the particle, but acts opposite to its direction of motion. The third term represents the random force $\mathbf{f}(t)$, whose time average is zero. The random force contains details of the "fast" dynamics. It is the resultant of a large number of microscopic collisions, and so follows a Gaussian distribution by virtue of the central limit theorem [11,23,26]. These collisions can generally be assumed to be uncorrelated on time scales much longer than the collision time, and can be regarded as a sequence of independent impulses. Since the Brownian particle attains thermal equilibrium with the fluid particles over a sufficiently long time, these considerations dictate the second moment of the random force given by

$$\langle f_n(t) f_m(t') \rangle = 2k_B T \zeta \delta(t - t') \delta_{nm}. \tag{7.31}$$

The delta function indicates that the force at t is uncorrelated with the force at t'. Since the process is also assumed to be Gaussian, all cumulants higher than the second are zero.

Equation (7.31) is the simplest form of the fluctuation dissipation theorem [22]. It states that the dissipation due to frictional forces and the fluctuations due to random forces are related to one another since they have the same microscopic origin.

For long enough length and time scales, the Langevin equation describes the Brownian motion of the centre of mass of the polymer. In contrast, the chain-like dynamical properties are characterized by length- and time scales that are longer than the dynamics characteristic of the monomers, but that are shorter than the Brownian dynamics characteristic of the centre of mass. The dynamical behaviour of the polymer in this intermediate regime is independent of the chemistry of the chain. The intermediate time regime, therefore, describes the dynamical aspects of the universal chain-like properties [21, 31, 36, 39]. The simplest model that describes low frequency global features of the polymer dynamics through a few phenomenological parameters is referred to as the Rouse model [31]. This model considers the chain to be a Gaussian equivalent chain with N beads and Kuhn length b. In the Rouse model, a polymer is represented as a bead and spring chain as shown in Fig. 7.2. Apart from the inertial force, the frictional force and the random force, the ith bead also experiences an elastic force from its immediate neighbours. The equation of motion for the ith bead is given by

$$m\ddot{\mathbf{r}}_n(t) = -\zeta_b \dot{\mathbf{r}}_n(t) + \mathbf{f}_n(t) - \frac{3k_B T}{b^2} \sum_{n=1}^{N} A_{nm} \mathbf{r}_m(t), \tag{7.32}$$

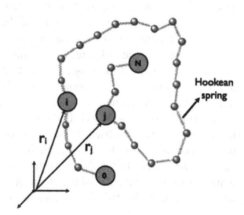

Fig. 7.2 Bead and spring Rouse chain

where ζ_b is the bead friction coefficient, which is related to the random force $\mathbf{f}_n$, as in the case of the Brownian motion. The last term of the above equation accounts for the elastic force, with A_{nm} being the Rouse matrix, defined as

$$A_{nm} = 2\delta_{nm} - \delta_{n,m+1} - \delta_{n,m-1}. \tag{7.33}$$

Since polymers are large molecules, the dynamics of polymers in solutions is dominated by viscous effects. In the over-damped limit, the inertial term in (7.32) can be ignored. Therefore, the equation of motion in the continuum limit is given by

$$\zeta \dot{\mathbf{r}}_n(t) = \frac{3k_B T}{b^2} \frac{\partial^2 \mathbf{r}(n,t)}{\partial n^2} + \mathbf{f}_n(t). \tag{7.34}$$

In general, force acting on any point along the chain backbone can cause motion of the fluid around it, which in turn can affect the velocity of the other segments. This interaction, which is mediated by the motion of the solvent fluid, is called the hydrodynamic interaction. A more general expression for the dynamics of polymers is given by

$$\zeta \dot{\mathbf{r}}_n(t) = -\int_0^N dn' D[\mathbf{r}_n - \mathbf{r}_{n'}] \frac{\delta H}{\delta \mathbf{r}_n'} + \mathbf{f}_n(t), \tag{7.35}$$

where H is the flexible chain Hamiltonian given by (7.11). The non-local hydrodynamic interaction is accounted for by the kernel $D[\mathbf{r}_n - \mathbf{r}_{n'}]$. The Rouse model neglects the hydrodynamic interactions. It assumes that the velocity at any point along the chain backbone is determined only by the force acting on it. The kernel is, therefore, given by $\delta[\mathbf{r}_n - \mathbf{r}_{n'}]$. The model that accounts for non-local hydrodynamic interactions in pre-averaged fashion is referred to as the Zimm model [39].

A standard way of treating the Rouse model is to decompose the motion of the polymer into the normal co-ordinates such that each mode is capable of independent motion. The decomposition into independent modes is done by identifying the normal mode variable $q = 2\pi p/N$, where $p = 0, 1, 2, 3 \cdots$ are the Rouse mode variables. A small q explores large length scales inside the chain; the $p = 0$ mode, therefore, represents the motion of the centre of mass of the chain. Several dynamical quantities of interest can be expressed in terms of these normal modes [7]. However, a simple estimate of the diffusion coefficients and the relaxation times can be made from the following considerations: according to Einstein's relation, the diffusion coefficient of a Brownian particle is given by $D = k_B T/\zeta$; the relaxation time can be estimated by determining the time required for the particle to move a distance of its size, $\tau \sim R^2/D$. In the Rouse model, each bead has the friction coefficient of ζ. So the total friction coefficient of the chain is $\zeta_R = N\zeta$. Therefore, the diffusion coefficient of the Rouse chain is $D_R = k_B T/N\zeta$ and the relaxation time is $\tau_R \sim N^2 b^2 \zeta/k_B T$. In the Zimm model, the hydrodynamic interactions couple the dynamics of different parts of the chain. As a result, the Zimm chain moves like a solid object with the friction coefficient governed by the Stoke's law, $\zeta_Z \sim \eta R$, where η is the solvent viscosity. The diffusion coefficient and the relaxation time of the Zimm chain are given by $D_Z = k_B T / \eta R$ and $\tau_Z \sim N^{3/2} b^3 \eta / k_B T$, respectively.

In the next section, the estimation of a few dynamical quantities under flow fields is discussed by modifying the Rouse model to account for the effects of flow.

7.9 Polymers Under Flow

Polymers are known to deform under various kinds of flow, the extent of deformation depending strongly on the nature of flow [1, 4, 8–10, 19, 30, 35]. A pure elongational flow produces large changes in the size of the polymer coil. A pure rotational flow, on the other hand, rotates the chain without inducing deformation. Most practical flows consist of a mixture of elongational and rotational components, so the dynamics of a chain in such flows depends on the relative magnitudes of these two components. In a simple shear flow, the magnitudes of the rotational and elongational components are equal; the stretched state produced by the elongational component is destabilized by the rotational component. Therefore, the extended state of polymers in shear flow is inherently unstable and undergoes large fluctuations [5]. The variation of the mean polymer size as a function of the flow rate and decay of the time correlation function for various values of the flow rate are two quantities that can be studied from statistical mechanical analysis [1, 4, 8–10, 19, 30, 35]. If the hydrodynamic interactions are ignored, the dynamics can be described by the Rouse model, supplemented with a term that accounts for the force field due to flow

$$\zeta \dot{\mathbf{r}}_n(t) = k \frac{\partial^2 \mathbf{r}(n,t)}{\partial n^2} + \dot{\gamma} \mathbf{A} \cdot \mathbf{r}_n(t) + \mathbf{f}_n(t), \qquad (7.36)$$

where $k = 3k_B T / b^2$ is the spring constant, $\dot{\gamma}$ is the flow rate and $\mathbf{A}$ is the velocity gradient tensor, which contains the details of the applied flow. The velocity gradient tensor for linear-mixed flow is given by

$$\mathbf{A} = \begin{pmatrix} 0 & 1 & 0 \\ \alpha & 0 & 0 \\ 0 & 0 & 0 \end{pmatrix},$$

where $\alpha = 0$ corresponds to simple shear, while $\alpha = 1$ and $\alpha = -1$ correspond to pure elongational and rotational flows, respectively. Equation (7.36) can be solved for the chain extension as a function of the flow rate. However, as discussed earlier, the Rouse chain is a bead and spring chain, consisting of Hookean springs that can be extended indefinitely. Real polymers show Hookean behaviour only for very small extensions. Hooke's law is expected to break down at large values of the applied flow. Equation (7.36) needs to account for the constraint of inextensibility.

For a discrete polymer model, Finitely Extensible Nonlinear Elasticity (FENE) ansatz is often used by writing the spring constant as $k = k'/\left(1 - l_i^2/l_m^2\right)$, where $k' = 3k_BT/b^2$, l_i is the instantaneous bond extension, and l_m is the maximum extension of the bond. Thus, the chain pays a high penalty for extension beyond its maximum length. A more practical ansatz preaverages the instantaneous bond extension to yield $k = k'/\left(1 - \langle l_i^2 \rangle/l_m^2\right)$. In the same vein, the following ansatz serves well for continuous chains [8–10]:

$$k = k' \left(\frac{1 - \langle \mathbf{R}^2 \rangle_0 / \langle \mathbf{R}^2 \rangle_m}{1 - \langle \mathbf{R}^2 \rangle / \langle \mathbf{R}^2 \rangle_m} \right), \qquad (7.37)$$

where $\langle \mathbf{R}^2 \rangle$ and $\langle \mathbf{R}^2 \rangle_0$ represent the mean square end-to-end distances in the presence and absence of flow. $\langle \mathbf{R}^2 \rangle_m$ is the maximum observed mean square end-to-end distance under flow.

In terms of the normal modes, $\mathbf{r}_p(t) = \frac{1}{N} \int_0^N dn \mathbf{r}_n(t) \cos(2p\pi n/N)$, the above equations can be rewritten as

$$\zeta_p \frac{\partial \mathbf{r}_p(t)}{\partial t} = k_p \mathbf{r}_p(t) + \dot{\gamma} \mathbf{r}_p(t) + \mathbf{f}_p(t), \qquad (7.38)$$

where $k_p = p^2\pi^2 k/N$, $\zeta_0 = N\zeta$ and $\zeta_{p \neq 0} = 2N\zeta$. The above expression can be solved for the chain extension under steady state to yield

$$\langle \mathbf{R}^2 \rangle = \frac{8}{N} \sum_{p=1, \text{odd}} \left[\frac{1}{k_p} + \frac{\dot{\gamma}^2 \zeta^2 (1+\alpha)^2}{2k_p (k_p^2 - \dot{\gamma}^2 \zeta^2 \alpha)} \right]. \qquad (7.39)$$

If $\tau = \zeta N^2 b^2/3\pi^2 k_B T$ is defined as the longest relaxation time and $Wi = \dot{\gamma}\tau$ as the Weissenberg number, the above equation can be rewritten as

$$z = \beta \left[\frac{(1-z)}{(1-\beta)} + \frac{4}{3\pi^2} \left(\frac{1-z}{1-\beta} \right)^3 Wi^2 (1+\alpha)^2 S \right], \qquad (7.40)$$

where $z = \langle \mathbf{R}^2 \rangle / \langle \mathbf{R}^2 \rangle_m$, $\beta = \langle \mathbf{R}^2 \rangle / \langle \mathbf{R}^2 \rangle_m$, and S is given by

$$S = -\frac{\pi^2}{8\alpha Wi^2} \left(\frac{1-\beta}{1-z} \right)^2$$
$$+ \frac{\pi}{8(\alpha Wi^2)^{5/4}} \left(\frac{1-\beta}{1-z} \right)^{5/2} (\tan(a\pi/2) + \tanh(a\pi/2)), \qquad (7.41)$$

where $a = (\alpha Wi^2)^{1/4}((1-\beta)/(1-z))^{1/2}$.

If the mean fractional extension of the chain is defined as $x = \sqrt{z}$, the results can be presented in terms of Figs. 7.3 and 7.4, which show the variation of the mean fractional extension as a function of the Weissenberg number Wi [10].

Figure 7.3 compares the theoretical result for the mean fraction extension as a function of Wi with results from the Brownian dynamics simulations of

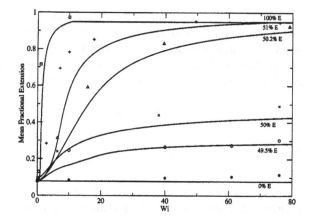

Fig. 7.3 Mean fractional extension, x, versus the Weissenberg number, Wi, for different fixed values of α. The *full lines* are the theoretical curves calculated from (7.40) and (7.41). The symbols represent data from simulations of [19] with the following meanings: *open squares*, $100\%E$; *plus signs*, $51\%E$; *crosses*, $50\%E$, *open circles*, $49.5\%E$; *asterisks*, $0\%E$. This figure has been reproduced from [10]

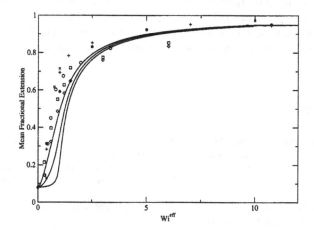

Fig. 7.4 Mean fractional extension, x, versus the effective Weissenberg number, $Wi^{eff} = \sqrt{\alpha}Wi$ for $\alpha > 0$. The *full lines* are the theoretical curves calculated from (7.40) and (7.41). The symbols represent data from simulations of [19]. This figure has been reproduced from [10]

Chu et al. [19] for six different values of α represented in terms of $\%E = 50(1+\alpha)$. In a pure extensional flow ($100\%E$), the chain undergoes a sudden increase in its size at very small values of Wi. In a pure rotational flow ($0\%E$), the chain only rotates and does not extend. In the case of simple shear, the presence of a large number of fluctuations results in 50% full extension. In the latter case, our theoretical curves reproduce the trend seen in single molecule experiments on DNA under shear [1, 35]. Figure 7.4 represents the variation of mean extension as a function of $Wi^{eff} = \sqrt{\alpha}Wi$. A significant degree of data collapse suggests that Wi^{eff} is a reasonable scaling variable for $\alpha > 0$. Our theoretical predictions compare well with simulation and experimental data on polymers under flow. The neglect of the hydrodynamic interactions leads to deviations at low Weissenberg numbers. The finitely extensible Zimm model can provide a better comparison at low Weissenberg numbers.

7.10 Conclusions

In this chapter, a unified description of ideal polymers is presented in terms of Gaussian equivalent chains. The static properties of Gaussian equivalent chains, like the size and its dependence on solvent quality and temperature, are discussed in terms of simple scaling arguments and theories based on statistical mechanics. The non-linearity in the force–extension relation for real polymers is obtained from the concept of tension blobs. The diffusion coefficient and the relaxation time of a polymer are calculated from the Rouse model. In the presence of the hydrodynamic interactions, the dynamic quantities of interest are estimated from the Zimm model. The behaviour of polymers under flow is presented in terms of the finitely extensible Rouse model, which is shown to provide a reasonably accurate description of polymers in weak extensional, rotational, simple shear and linear-mixed flows.

References

1. Babcock HP, Smith DE, Hur JS et al (2000) Relating the microscopic and macroscopic response of a polymeric fluid in a shearing flow. Phys Rev Lett 85:2018–2021
2. Birshtein TM, Ptitsyn OB (1963) Conformations of macromolecules. Interscience, New York
3. Chandrasekhar S (1969) Stochastic problems in physics and astronomy. Rev Mod Phys 15:1–89
4. de Gennes PG (1974) Coil-stretch transition of dilute flexible polymers under ultrahigh velocity gradients. J Chem Phys 60:5030
5. de Gennes PG (1979) Scaling concepts in polymer physics. Cornell University Press, Ithaca
6. Doi M (1996) Introduction to polymer physics. Clarendon Press, Oxford
7. Doi M, Edwards SF (1986) The theory of polymer dynamics. Oxford University Press, Oxford
8. Dua A, Cherayil BJ (2000) Chain dynamics in steady shear flow. J Chem Phys 112:8707
9. Dua A, Cherayil BJ (2000) Effects of stiffness on the flow behavior of polymers. J Chem Phys 113:10776
10. Dua A, Cherayil BJ (2003) Polymer dynamics in linear mixed flow. J Chem Phys 119:5696
11. Feller W (1971) An introduction to probability theory and its applications, Vol 2, Wiley, New York
12. Feynman RP, Hibbs AR (1966) Quantum mechanics and path integrals. Springer, Berlin
13. Flory PJ (1953) Principles of polymer chemistry. Cornell University Press, Ithaca, New York
14. Flory PJ (1969) Statistical mechanics of chain molecules. Interscience, New York
15. Freed K (1971) Weiner integrals and models of stiff polymers. J Chem Phys 54:1453
16. Freed K (1972) Functional integral and polymer statistics. Adv Chem Phys 22:1–120
17. Freed K (1987) Renormalization group theory of macromolecules. Wiley, New York

18. Harris RA, Hearst JE (1966) On polymer dynamics. J Chem Phys 44:2595
19. Hur JS, Shaqfeh ESG, Babcock HP et al (2002) Dynamics and configurational fluctuations of single DNA molecules in linear mixed flows. Phys Rev E 66:011915
20. Ito K, Mckean HP (1965) Diffusion processes and their simple paths. Springer, Berlin
21. Iwata K (1971) Irreversible statistical mechanics of polymer chains, I. Fokker-Planck diffusion equation. J Chem Phys 54:12
22. Kadanoff L, Martin P (1963) Hydrodynamic equations and correlation function. Ann Phys 24:419
23. Kubo R, Toda M, Hashitsume N (1984) Statistical physics II: Nonequilibrium statistical mechanics. Springer, Berlin
24. Lagowski JB, Noolandi J, Nickel B (1991) Stiff chain model–functional integral approach. J Chem Phys 95:1266
25. Langevin P (1908) On the theory of Brownian motion. C R Acad Sci, Paris 146:530
26. Ma SK (1985) Statistical mechanics. World Scientific, Singapore
27. Mandelbrot B (1982) The fractal geometry of nature. W.H. Freeman and Company, New York
28. Mortensen RE (1969) Mathematical problems of modeling stochastic nonlinear dynamic systems. J Stat Phys 1:271
29. Pearson K (1905) The problem of the random walk. Nature 77:294
30. Perkins TT, Smith DE, Chu S (1997) Single polymer dynamics in an elongational flow. Science 276:2016–2021
31. Rouse PE (1953) A theory of the linear viscoelastic properties of dilute solutions of coiling polymers. J Chem Phys 21:1272
32. Rubenstein M, Colby RH (2003) Polymer physics. Oxford University Press, Oxford
33. Saito N, Takahashi K, Yunoki Y (1967) The statistical mechanical theory of stiff chains. J Phys Soc Jpn 22:219–226
34. Schulman LS (1981) Techniques and applications of path integration. Wiley, New York
35. Smith DE, Babcock HP, Chu S (1999) Single polymer dynamics in steady shear flow. Science 283:1724–1727
36. Verdier PH, Stockmayer WH (1962) Monte Carlo calculation on the dynamics of polymers in dilute solution. J Chem Phys 36:227
37. Volkenstein MV (1963) Configurational statistics of polymer chains. Interscience, New York
38. Yamakawa H (1971) Modern theory of polymer solutions. Harper and Row, New York
39. Zimm BH (1962) Dynamics of polymer molecules in dilute solution: Viscoelasticity, flow birefringence and dielectric loss. J Chem Phys 36:227

Chapter 8
Polymer Rheology

P. Sunthar

Abstract This chapter concerns the flow behavior of polymeric liquids. This is a short introduction to the variety of behaviours observed in these complex fluids. It is addressed at the level of a graduate student, who has had exposure to basic fluid mechanics. The contents provide only a broad overview and some simple physical explanations to explain the phenomena. For detailed study and applications to industrial contexts, the reader is referred to more detailed treatises on the subject. The chapter is organised into three major sections.

8.1 Phenomenology

Interesting behavior of a material is almost always described by first observing and extracting the qualitative features of the variety of phenomena exhibited. To a novice, this is the place to start. Some questions to ask would be: What makes this phenomenon different? How to depict it in terms of a simple model? Is there a "law" that can describe the behavior? Are there other phenomena that obey similar laws? What role has this played in the state of the universe? Can it be employed for the betterment of quality of life? What are the consequences of this behavior to processes that manipulate or use the material?

Some of these are simple questions, answers to which may not solve any urgent and present problem. However, it puts it in a larger context. Often, when solving difficult problems in rheology using advanced methods and tools, one tends to miss simple laws exhibited by the materials in reality, knowing which may help in getting an intuitive feel of the solution, and thereby a faster approach to the answer.

P. Sunthar (✉)
Department of Chemical Engineering, Indian Institute of Technology Bombay,
Mumbai 400076, India
e-mail: p.sunthar@gmail.com

J.M. Krishnan et al. (eds.), *Rheology of Complex Fluids*,
DOI 10.1007/978-1-4419-6494-6_8, © Springer Science+Business Media, LLC 2010

8.1.1 What are Polymeric Liquids?

Polymeric liquids are precisely what the name states: namely, they are like the liquids we know that flow and have as constituents, some or all long chain molecules or polymers. The conventional definition used to define *simple liquids* that they do not support shear stress at rest cannot be used to define the liquid state of a polymeric liquid. This is because polymeric liquids, like most other liquids described in this book, are *complex fluids*: they exhibit both liquid- and solid-like behaviours, and some of their dynamic properties may not be thermodynamic constants but some effective constant dependent on the history of forces acting on it. An example is a "viscosity", which is a function of shear rate, or a "viscosity" that changes with time.

8.1.1.1 Chemical Nature

The common feature of all polymeric liquids considered in this chapter is that they have long chains of monomers joined by chemical bonds. They could even be oligomers or very long chain polymers. Most of them have large molecular weights, more than a 1,000 and up to about 10^9. Many materials we use today are polymers or blends of polymers with other materials. In some stage of their processing, most of them were in one liquid state or other: as solutions or pure molten forms.

8.1.1.2 Physical Nature

The chief physical property of a polymer that distinguishes it from other fluids that exhibit complex behavior is the linearity of the chain. It is not the high molecular weight that leads to the peculiar phenomena but that it is arranged *linearly*: The length along the chain is much larger that the other dimensions of the molecule. For example, a suspension of polystyrene beads (in solid state) may have high molecular weight per bead. But it may exhibit rheological properties of a suspension rather than a polymer solution. This is because for most properties that we measure, such as the viscosity, it is the beads' spherical diameter that matters, and not the molecular weight per bead. There is no linear structure that is exposed to these measurements. On the other hand, a solution of polystyrene molecules in cyclohexane is a polymeric liquid.

What is the difference between a polymeric liquid to us and a huge bowl of noodles to a Giant?

A huge bowl of noodles is very much like several molecules of polymers put together. Just as we "see" polymers, so the bowl with these noodles would appear to a Giant. The noodles have the necessary linearity, they are flexible like the polymers, and as a whole can take the shape of the container they are put into. Can we then use a model bowl full of noodles to understand the behavior of polymeric liquids? Not entirely! The key difference between the two systems is temperature or random motion. Many of the properties of polymeric liquids we observe, such as that they flow, are due to the *random linear translating motion* of their constituents, similar to simple liquids. It is the random motion along the length of the chain and the resulting influence on the other constituents of the material, such as the polymer itself, solvent or other polymers, that leads to the *defining properties* of polymeric liquids.

"States" of Polymeric Liquids

The simplest definition (though not a theoretically simpler model to handle) of a polymeric liquid state is the *molten state*. Take a pure polymer (with no other additives) to a high enough temperature that it is molten. This is like a noodle-state, except that the polymers are in continuous motion. The other extreme of this state is when small amounts of a polymer are added as an additive (solute) to a solvent. This state is called the *dilute solution*. This is similar to putting one polymer chain in a sea of solvent. It is equivalent to one polymer because the solution is so dilute that the motion of one polymer does not influence the other polymers in solution. Parts of the polymer's motion influence other parts of the same chain. Between these two states of pure and dilute, we can have a range of proportions of the polymer and the other component. Increasing the concentration of the additive polymer from dilute values, we get to the *semi-dilute* region where the polymers are distributed in the solution such that they just begin to "touch" each other. Further increase in concentration leads to the *concentrated-solution* where there are significant overlaps and *entanglements* (like in an entangled noodle soup). The molten state is like a dry noodle full of entanglements.

8.1.2 Who Needs to Study Flow of Polymeric Liquids?

The primary motivation to study polymeric fluids comes from industrial and commercial applications: polymer processing and consumer products. These fields have provided (and continue to provide) several problems that are addressed by a wide variety of engineers and physicists to the extent that they have even earned a Nobel Prize in Physics for Pierre-Gilles de Gennes in 1991.

The primary question is how the flow is modified in the presence of these long chain macromolecules. In the polymer processing field, one needs to know, as in any fluid mechanics problem, how to control the flow and stability of these liquids as they pass through various mixers, extruders, moulds, spinnerets, etc. For consumer products, on the other hand, we would like to able to control the feel, rates, shapes, etc. of liquids by having additives, which in many cases are polymers.

8.1.3 What is Polymer Rheology?

Clearly, industrial flows are complex not only because the geometries are complex but also because the constituents are not usually simple. We have several components in a shampoo, performing various actions. Molecular weight distribution of a polymer is another level of complexity, as polymers are rarely synthesised in a sharp monodisperse population.

This brings in the need to study the behavior of polymeric liquid in simple flows and for simple systems, with the hope that the knowledge gained can be appropriately used in a complex flow pattern. The word *Rheology* is defined as the science of deformation and flow, and was coined by Prof Bingham in the 1920s [1]. Rheology involves measurements in controlled flow, mainly the viscometric flow in which the velocity gradients are nearly uniform in space. In these simple flows, there is an applied force where the velocity (or the equivalent shear rate) is measured, or vice versa. They are called *viscometric* as they are used to define an effective shear viscosity η from the measurements

$$\eta = \frac{\sigma_{xy}}{\dot{\gamma}}, \tag{8.1}$$

where σ_{xy} is the shear stress (measured or applied) and $\dot{\gamma}$ is the shear rate (applied or measured). Viscosity is measured in Pascal second (Pa s).

Rheology is not just about viscosity, but also about another important property, namely the elasticity. Complex fluids also exhibit elastic behavior. Akin to the viscosity defined above being similar to the definition of a Newtonian viscosity, the elasticity of a complex material can be defined similar to its idealised counterpart, the Hookean solid. The modulus of elasticity is defined as

$$G = \frac{\sigma_{xy}}{\gamma}, \tag{8.2}$$

where γ is called the strain or the angle of the shearing deformation. G is measured in Pascal (Pa). G is one of the elastic moduli, known as the storage modulus, as it is related to the amount of recoverable energy stored by the deformation. G for most polymeric fluids is in the range 10–10^4 Pa, which is much smaller than that of solids ($>10^{10}$ Pa). This is why complex fluids, of which polymeric fluids form a major part, are also known as *soft matter*, i.e. materials that exhibit weak elastic properties.

Rheological measurements on polymers can reveal the variety of behaviours exhibited even in simple flows. Even when a theoretical model of the reason for the behavior is not known, rheological measurements provide useful insights into practising engineers on how to control the flow (η) and feel (G) of polymeric liquids. In the following section, we sample a few defining behaviours of various properties in commonly encountered flows. We present a broad overview of the variety of phenomena observed in polymeric liquids and how the rheological characterisation of the liquids can be made.

8.1.4 Visual and Measurable Phenomena

Some of the striking visual phenomena are associated with flow behavior of polymeric liquids. These are best seen in recorded videos. We describe a few of them here and provide links to resources where they may be viewed. There are many more phenomena discussed and illustrated in [3].

8.1.4.1 Weissenberg Rod Climbing Effect

When a liquid is stirred using a cylindrical rod, the liquid that wets the rod begins to "climb" up the rod and the interface with the surrounding air assumes a steady shape dangling from the rod, so long as there is a continuous rotation. In contrast in a Newtonian liquid, there is a dip in the surface of the liquid near the rod. Rod climbing is exhibited by liquids that show a normal stress difference. In Newtonian liquids, the normal stresses (pressures) are isotropic even in flow, whereas polymeric liquids, upon application of shear flow, begin to develop normal stress differences between the flow (τ_{xx}) and flow-gradient directions (τ_{yy}).

8.1.4.2 Extrudate or Die Swell

This phenomenon is observed when polymeric melts are extruded through a die. The diameter of liquid as it exits a circular die can be three times larger than the diameter of the die, whereas in the case of Newtonian fluids it is just about 10% higher in the low Reynolds number limit. One of the important reasons for this phenomenon is again the normal stress difference induced by the shear flow in the die. As the fluid exits the die to form a free surface with the surrounding air, the accumulated stress difference tends to push the fluid in the gradient direction.

8.1.4.3 Contraction Flow

Sudden contraction in the confining geometry leads to very different streamline patterns in polymeric liquids. In Newtonian liquids at low Reynolds number, no secondary flows are observed, whereas in polymeric liquids, including in dilute polymer solutions, different patterns of secondary flow are observed. These include large vortices and other instabilities. These flows are undesirable in many situations in polymer processing as they lead to stagnation and improper mixing of the fluid in the vortices.

8.1.4.4 Tubeless Siphon

In a typical siphoning experiment, a tube filled with liquid drains a container containing the liquid at a lower pressure, even though the tube goes higher than the liquid surface. When the tube is lifted off the surface of the liquid, the flow immediately stops. But this is the case in Newtonian liquids. In the case of polymeric liquids, the liquid continues to flow with a free surface with the air without the tube, as the tube is taken off the surface.

8.1.4.5 Elastic Recoil

A "sheet" of polymeric liquids pouring down from a vessel can be literally cut with a pair of scissors. Very similar to a sheet of elastic solid, the top portion of the cut liquid recoils back into the jar.

8.1.4.6 Turbulent Drag Reduction

In most of the phenomena discussed so far the concentration of the polymer is about 0.1% or higher, and the viscosities of such systems are usually large that it is not common to encounter large Reynolds number flows that lead to turbulence. In smaller concentrations of about 0.01% where the solution viscosity is not significantly enhanced above the solvent's viscosity, turbulence can be easily observed. The interesting feature of such turbulent flows, at least in pipe geometries, is that the turbulent friction on the walls is significantly less, up to nearly five times. This phenomenon has been used in transportation of liquids and in fire-fighting equipment.

8.1.5 Relaxation Time and Dimensionless Numbers

One of the simplest and most important characters of polymeric liquids is the existence of an observable microscopic time scale. For regular liquids, the time scales of molecular motion are in the order of 10^{-15} s, associated with molecular translation. In polymeric liquids, apart from this small time scale, there is an important time scale associated with large scale motions of the whole polymer itself, in the liquid they are suspended in (solutions or melts). This could be from microseconds to minutes. Since many visually observable and processing time scales are of similar order, the ratio of these time scales becomes important. The large scale microscopic motions are usually associated with the elastic character of the polymeric liquids. In the chapter on polymer physics, there is a discussion of the relaxation times. The relaxation time is the time associated with large scale motion (or changes) in the structure of the polymer; we denote this time scale by λ.

The microscopic time scale should be compared with the macroscopic flow time scales. The macroscopic time scales arise from two origins. One is simply the *kinematic* local rate of stretching of the fluid packet (*strain rate*). This is measured by the local shear rate $\dot{\gamma}$ for shearing flows or the local elongation rate $\dot{\varepsilon}$ for extensional flows. The other is a *dynamic* time scale associated with the motion of the fluid packets themselves. Examples are the time it takes for a fluid packet to transverse a geometry or a section, pulsatile flow, etc. We denote this time scale by t_d. Except for viscometric flows, the macroscopic time scales may not be known a priori, and have to be determined as part of the solution. For example, the nature of the fluid's viscosity could alter the local shear rate $\dot{\gamma}$, or the time it spends in a particular section t_d. To know this dependence, we need to solve the fluid dynamics equations in the given geometry.

8.1.5.1 Weissenberg Number

The ratio of the microscopic time scale to the local strain rate is called as the Weissenberg number:

$$\text{Wi} = \lambda\dot{\gamma} \quad \text{or} \quad \lambda\dot{\varepsilon}. \tag{8.3}$$

Note that the strain rate is the inverse of the kinematic time scale. Flows in which the Wi are small, $\text{Wi} \ll 1$, are those in which elastic effects are negligible. Most of the flow effects are seen around $\text{Wi} \sim \mathscr{O}(1)$. For large Wi, $\text{Wi} \gg 1$, the liquid behaves almost like an elastic solid.

The Weissenberg number is used only in situations where there is a homogeneous stretching of the fluid packet in the flow. That is, the strain rates are uniform in space and time. Such a flow is encountered only in viscometric flows and theoretical analysis, as it is hardly observed in any practical application.

8.1.5.2 Deborah Number

In most practical applications, the fluid packets undergo a non-uniform stretch history. This means that they could have been subjected to various strain rates at various times in their motion. Therefore, no unique strain rate can be associated with the flow. In these cases, it is customary to refer to the Deborah number defined as the ratio of the polymeric time scale to the dynamic or flow time scale:

$$\text{De} = \frac{\lambda}{t_d}. \tag{8.4}$$

For small De, $\text{De} \ll 1$, the polymer relaxes much faster than the fluid packet traverses a characteristic distance, and so the fluid packet is said to have "no memory of its state" a few t_d back. On the other hand, for $\text{De} \sim \mathscr{O}(1)$, the polymer has not sufficiently relaxed and the state a few t_d back can influence the motion of the packet now (because this can affect the local viscosity and hence the dynamics).

Table 8.1 Scaling of the relaxation time λ of the polymeric liquid with molecular weight M for different classes of polymeric liquids [10]

Class	Scaling
Dilute solution in poor solvent	$\lambda \sim M^{1.0}$
Dilute solution in θ-conditions	$\lambda \sim M^{1.5}$
Dilute solution in good solvent	$\lambda \sim M^{1.8}$
Semi-dilute solution	$\lambda_{chain} \sim M^2$
Entangled melts	$\lambda_{rep} \sim M^{3.4}$

8.1.5.3 Relaxation Time Dependence on Molecular Weight

Since the time scale motion of the molecule (i.e. its relaxation time) depends on the linear size of the molecule, the molecular weight of a polymer has a direct bearing on the relaxation times. The dependence on molecular weight is *not absolute*. That is, we cannot say that for a given molecular weight two different polymers will have the same relaxation time. It is only a scaling dependence for a class of polymeric liquids. (We can only say that the relaxation time scales with molecular weight power some exponent). This dependence is summarised in Table 8.1. The scaling given for semi-dilute and entangled melts is only indicative of the longest relaxation time; there are several relaxation processes in these systems, and the way experimental data is interpreted from measurements carried out at various temperatures (see Sect. 8.7 of Ref. [10] for details).

8.1.6 Linear Viscoelastic Properties

In general, the elastic nature of a material is associated with some characteristic equilibrium microstructure in the material. When this microstructure is disturbed (deformed), thermodynamic forces tend to restore the equilibrium. The energy associated with this restoration process is the elastic energy. Polymeric liquids have a microstructure that is like springs representing the linear chain. Restoration of these springs to their equilibrium state is through the elastic energy that is "stored" during the deformation process. But polymeric fluids are not ideal elastic materials, and they also have a dissipative reaction to deformation, which is the viscous dissipation. For small deformations, the response of the system is linear, meaning that the response is additive: the effect of the sum of two small deformations is equal to the sum of the two individual responses. Linear viscoelasticity was introduced in the chapter on Non-Newtonian Fluids.

Linear viscoelastic properties are associated with near equilibrium measurements of the system, that is, the configuration of polymers is not removed far away from their equilibrium structures. Most of the models described in the "Polymer Physics" chapter deal with such a situation. A study of linear viscoelastic properties can reveal information about the microscopic structure of polymeric liquids. The term rheology is used by physicists to usually refer to the linear response and by engineers to refer

Table 8.2 Linear viscoelastic properties of common liquids. Values are typical order of magnitude approximations [1]

Liquid	Viscosity η (Pa s)	Relaxation time λ (s)	Modulus G (Pa)
Water	10^{-3}	10^{-12}	10^9
Oil	0.1	10^{-9}	10^8
A polymer solution	1	0.1	10
A polymer melt	10^5	10	10^4
Glass	$>10^{15}$	10^5	$>10^{10}$

to the large deformation (shear or elongational) or non-linear state. We will first present aspects of the linear response before detailing the non-linear responses in the next sections.

An important point to note here is that it is just not polymeric liquids that show elastic behavior, but *all* liquids do, at sufficiently small time scales. Typical values of relaxation times and the elastic moduli for various liquids are given in Table 8.2.

Commonly used tests to study the linear response are:

Oscillatory: Controlled stress/strain is applied in a small amplitude oscillatory motion and the response of the strain/stress is measured.

Stress Relaxation: A constant strain is applied and the decay of the stresses to the equilibrium value is studied.

Creep: A constant stress is applied and the deformation response is measured.

Though all of the above tests can also be carried out in the non-linear regime, as the limits of linear regime are not known a priori, a sequence of tests is carried out to ensure linear response.

8.1.6.1 Zero-Shear Rate Viscosity

The zero-shear rate viscosity η_0 is the viscosity of the liquid obtained in the limit of shear rate tending to zero. Though the name suggests that it is a shear viscosity, it is still in the linear response regime because the shear rate is approaching zero. In practice, it is not possible to attain very low shear rates for many liquids owing to measurement difficulties. In these cases, the viscosity is obtained by extrapolating the viscosities obtained at accessible shear rates. The zero-shear rate viscosity is an important property to characterise the microstructure of a polymeric liquid.

For dilute solutions, since the polymer contribution to the total viscosity is usually small, it is useful to define an intrinsic viscosity at zero-shear rate as

$$[\eta]_0 \equiv \lim_{\dot{\gamma} \to 0} [\eta] \equiv \lim_{\dot{\gamma} \to 0} \lim_{c \to 0} \frac{\eta - \eta_s}{c \eta_s}. \tag{8.5}$$

The intrinsic viscosity scales as

$$[\eta]_0 \sim \frac{\lambda}{M},\tag{8.6}$$

where the scaling of λ is given in Table 8.1 for various types of dilute solutions.

In semi-dilute regime, one is more interested in the scaling of the viscosity with concentration. The usual way to report viscosity is through the specific viscosity (and not the intrinsic viscosity)

$$\eta_{sp0} = \eta_0 - \eta_s.\tag{8.7}$$

The specific viscosity scales linearly with concentration in the dilute regime. In the semi-dilute regime, under θ-conditions

$$\eta_{sp0} \sim c^2,\tag{8.8}$$

and in the concentrated (and melt) regime,

$$\eta_{sp0} \sim c^{14/3}.\tag{8.9}$$

The scaling with respect to concentration in good solvents is weaker to $\sim c^{1.3}$ in semi-dilute and $\sim c^{3.7}$ in the concentrated states. More details of this scaling can be found in Ref. [10]. The scaling behavior is summarised in Fig. 8.1.

8.1.6.2 Oscillatory Response

The typical response of a polymeric melt to an oscillatory experiment is shown in Fig. 8.2. The symbol used here is the complex modulus G^*: G' for the storage component (real) and G'' for the loss component (imaginary), as defined in the Non-Newtonian Fluids chapter. G' represents the characteristic elastic modulus of the system and G'' measures the viscous response. At high frequencies, the response is glassy (which is typically seen at temperatures around glass transition). The dominant elastic response is seen in the rubbery region where the storage modulus shows a plateau. The plateau region is clear and pronounced in higher molecular weight polymers (with entanglements) in the concentrated solution or the melt states, as shown in Fig. 8.3 [1]. In this region, the storage modulus (elastic) is always greater than the loss (viscous) modulus. The value of G' at the plateau is known as the plateau modulus G_N^0 and is an important property in understanding the dynamics of the polymers in the melt state.

The low-frequency response (or long-time response) is always viscous. The viscous regime behavior is characteristic of all materials (including solids) showing Maxwell behavior,

Fig. 8.1 Scaling of the specific viscosity η_{sp0} with concentration of polymers. The first transition denotes the semi-dilute regime and the second corresponds to the entangled regime

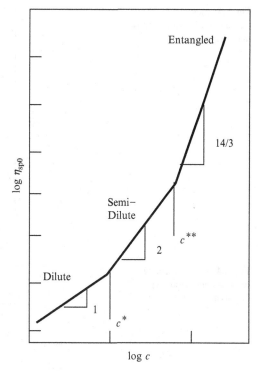

Fig. 8.2 Typical regimes in the complex modulus obtained using an oscillatory response of a polymeric liquid [6]

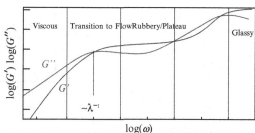

$$G' \approx G \lambda^2 \omega^2, \qquad (8.10)$$

$$G'' \approx \eta_0 \omega, \qquad (8.11)$$

where G is the elastic modulus (constant in the Maxwell model), and η_0 is the zero-shear rate viscosity. This behavior is also shown by polymeric liquids in the dilute and semi-dilute regimes. The characteristic relaxation time of the structured liquid can be obtained from the inverse of the frequency, where G' and G'' cross over in the flow transition regime.

$$\lambda = \frac{G'}{G'' \omega}, \qquad (8.12)$$

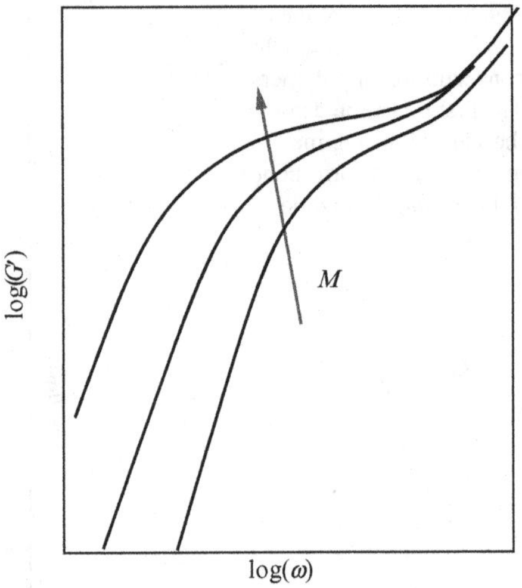

Fig. 8.3 Increase in the
plateau region with increase
in molecular weight

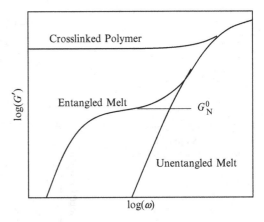

Fig. 8.4 Low frequency
response of various type
of polymers

where G' and G'' are measured in the viscous regime with the linear and quadratic
scaling, respectively. As seen before, increase in molecular weight increases the
relaxation time; so does the increase in concentration. Therefore with increasing
concentration or molecular weight of the polymer, the intersection of the two curves
shifts more and more towards the left (low frequencies). The low-frequency response of the loss modulus can also be used to obtain the zero-shear rate viscosity
η_0 from (8.11).

The low-frequency response close to the plateau region can also be used to distinguish the type of polymer melt. A schematic diagram of the plateau region behavior
of storage modulus G' for various types of polymers is shown in Fig. 8.4. Unen-
tangled melts do not have any plateau region, and directly make the transition to
the Maxwell region. Entangled melts show a plateau region. Cross-linked polymers
have a wide and predominant plateau region [6]. The transition to the Maxwell behavior is almost not seen due to limitations in observing every cycle, just as in solids.

The link between the plateau region and the cross-linking suggests that the entanglement acts like a kind of constraint (like the cross-links) to the motion of the polymer contour, leading to the plateau region.

8.1.6.3 Stress Relaxation

The relaxation of the stress modulus $G(t)$ in response to a step strain (for small strains) in the linear regime is equivalent to the oscillatory response $G^*(\omega)$, one being the Fourier transform of the other [8]. A schematic diagram of the stress relaxation is shown in Figs. 8.5 and 8.6 in linear and logarithmic scales, respectively [6]. The initial small time response of $G(t)$ is equivalent to the high-frequency response of $G^*(\omega)$, and the long-time response is equivalent to the low-frequency response. At small times, a polymeric substance shows a glassy behavior, which goes to the plateau region (seen clearly in the logarithmic scale in Fig. 8.6) and finally to the terminal viscous decay. Since G is related to the elasticity, the response can be understood as being highly elastic at small times, which decays and begins to "flow" at large times: given sufficiently long time, any material flows.

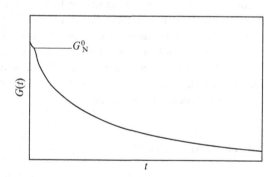

Fig. 8.5 Stress relaxation in response to step strain in linear scale

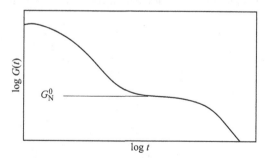

Fig. 8.6 Stress relaxation in response to step strain in logarithmic scale

8.1.7 Flow Viscosity and Normal Stress

So far we discussed the behavior of polymeric liquids slightly disturbed from equilibrium, in the linear regime. Here, we show the behavior in flowing systems that produce significant deviations from the equilibrium microstructure.

8.1.7.1 Shear Thinning

Most polymeric liquids have their effective viscosity reduced upon shearing. The viscosity defined in (8.1) is a decreasing function of the shear rate. Figure 8.7 shows the decrease in the viscosity for two polymeric liquids. The shear thinning is not so much pronounced in dilute solutions as it is in concentrated solutions and melts [1, 7].

A special class of polymers is known as *Living Polymers*, which are long linear structures formed from cylindrical liquid crystalline phases of micelles. They are called living polymers because they form and break along their length owing to thermodynamic and flow considerations. They are also known as *worm-like micelles*. At equilibrium, their behavior is similar to a high molecular weight concentrated solution of polymers: large viscosity and elastic modulus. However, upon shearing, they can break leading to the behavior exhibited by low molecular weight counterparts, aligning in the shearing direction. The viscosity reduction upon shearing is therefore very significant and sharp in these systems, as depicted schematically in Fig. 8.7. Examples of everyday use are shampoo and shower gels, which are very viscous at rest, but can be easily flown out of the container by gravitational forces (which are sufficient to overcome the viscosity after they begin to flow).

8.1.7.2 Normal Stresses

The normal stress difference is zero for a liquid that is isotropic. Polymeric liquids having microstructure can develop anisotropy in the orientation of the constituent

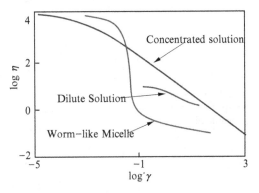

Fig. 8.7 Shear thinning of polymeric liquids is more pronounced in concentrated solutions and melts than in dilute solutions

polymers in flow, thereby leading to normal stress differences. The normal stress behavior shows a behavior similar to that of shear stress. The normal stress difference has two components,

$$N_1 = \tau_{xx} - \tau_{yy}, \tag{8.13}$$

$$N_2 = \tau_{yy} - \tau_{zz}, \tag{8.14}$$

where for planar Couette flow, x is the direction of flow, y is the direction of the gradient and z is the vorticity direction. Similar to the viscosity, which is a coefficient of the shear stress, we can define two coefficients for the normal stresses: the only difference is the denominator which is $\dot{\gamma}^2$, because it is the lowest power of the shear rate that the normal stresses depend on. The coefficients are called as first normal stress coefficients, Ψ_1 and Ψ_2, defined as:

$$\Psi_1 = \frac{N_1}{\dot{\gamma}^2}, \tag{8.15}$$

$$\Psi_2 = \frac{N_2}{\dot{\gamma}^2}. \tag{8.16}$$

The normal stress coefficients show shear thinning similar to the viscosity; however, for very large shear rates, the absolute value for the normal stress difference can become larger than the shear stress as shown in Fig. 8.8. Such a behavior is seen in concentrated solutions and in melts [1]. The second normal stress N_2 is usually zero for polymeric liquids.

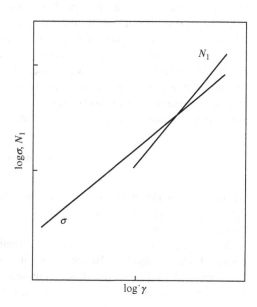

Fig. 8.8 Comparison of the absolute values of normal stresses with that of the shear stresses as a function of shear rate for concentrated solutions and melts

8.1.7.3 Elongational Flow Viscosity

Extensional or elongational flow is where the local kinematics dictates that the fluid element is stretched in one or more directions and compressed in others. Liquids being practically incompressible, the elongational stretch conserves the volume. Elongational flows are encountered in many situations. Though there are very few systems where it is purely extensional, there are several cases where there is significant elongation along with rotation of fluid elements, the two basic forms of fluid kinematics due to flow. Sudden changes in flow geometry are contraction or expansion, spinning of fibres, stagnation point flows, breakup of jets or drops, blow moulding, etc.

The simplest elongation is called as the uniaxial elongation, the velocity gradient tensor for which is given by

$$\nabla \mathbf{v} = \dot{\varepsilon} \begin{bmatrix} 1 & 0 & 0 \\ 0 & -\frac{1}{2} & 0 \\ 0 & 0 & -\frac{1}{2} \end{bmatrix}, \tag{8.17}$$

which corresponds to stretching along the x direction and equal compression along y and z directions. $\dot{\varepsilon}$ is called the elongation or extension rate. The elongational (or tensile) viscosity is defined as

$$\eta_E = \frac{\sigma_{xx} - \sigma_{yy}}{\dot{\varepsilon}} = \frac{\sigma_{xx} - \sigma_{zz}}{\dot{\varepsilon}}. \tag{8.18}$$

The elongational viscosity, like the shear viscosity, is a function of the shear rate. However, in the case of elongational flows, it is difficult to measure the steady-state value. In experiments, it is only possible to access a time-dependent value $\eta_E^+(t, \dot{\varepsilon})$, which is a transient elongational viscosity or more precisely the tensile stress growth coefficient [5]. The elongational viscosity is defined as the asymptotic value of this coefficient for large times, $t \to \infty$. The behavior of the transient growth coefficients for various elongation rates is shown in Fig. 8.9. The elongational viscosity abruptly increases to a high value at short time scales [2]. This phenomenon is called as strain hardening. The elongational strain is measured by the *Hencky strain*, defined as

$$\varepsilon = \dot{\varepsilon} t = \log \left(\frac{L(t)}{L_0} \right), \tag{8.19}$$

where L_0 is the initial length of an element along the stretch direction and $L(t)$ is the deformed length that grows as $L(t) = L_0 e^{\dot{\varepsilon} t}$. The elongational viscosity attains a maximum and then falls.

It is not possible to easily measure the terminal or asymptotic value of the viscosity. This is because for a Hencky strain of $\varepsilon = 7$, the elongation required is about 1,100 times the initial value. It is a convention therefore to report the maximum value of the transient tensile growth coefficient η_E^+ as a function of the strain rate

Fig. 8.9 Transient growth coefficient (transient elongational viscosity) for increasing strain rates $\dot{\varepsilon}$

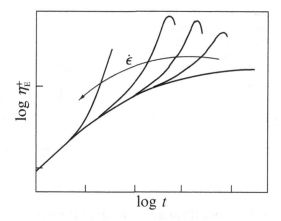

Fig. 8.10 Behavior of the elongational viscosity with elongation rate for various polymeric liquids. The elongational viscosity is not necessarily the steady-state value, and it could be the maximum or the value at the terminal Hencky strain

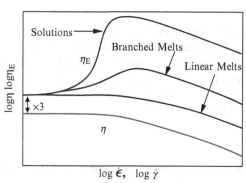

for practical applications. The "Steady values" of the viscosity reported are either the maximum values at a given strain rate, or the value of the coefficient at a given experimentally accessible Hencky strain. The behavior of this viscosity is shown in Fig. 8.10 for various polymeric liquids [1].

Another useful way to report the elongational viscosity is the Trouton ratio [2], which is defined as

$$T_R = \frac{\eta_E(\varepsilon)}{\eta(\sqrt{3}\,\dot{\varepsilon})}, \tag{8.20}$$

where the shear viscosity is measured at a shear rate $\dot{\gamma} = \sqrt{3}\,\dot{\varepsilon}$ (see further details in the chapter on Non-Newtonian Fluid Mechanics). For Newtonian (or inelastic) liquids, the elongational viscosity is three times the shear viscosity. For polymeric liquids, at low shear (and elongation) rates, the Trouton ratio T_R is always ≈ 3. The Trouton ratio plots, such as the one shown in Fig. 8.11, provide an indication of the extent of elongational viscosity effects in relation to the shear viscosity [1]. The most dramatic effects are in dilute solutions of polymers where, depending on the molecular weight of the polymer, the Trouton ratio can be several orders of magnitude higher than unity.

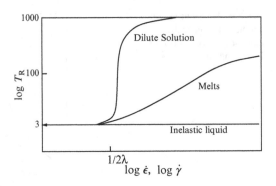

Fig. 8.11 Trouton ratios for various polymeric liquids. Dilute solutions show the most dramatic increase in the elongational viscosity

8.2 Modelling and Physical Interpretation

Modelling in polymer rheology is mainly of two types: phenomenological and molecular modelling. Some general phenomenological models have been discussed in the chapter on Non-Newtonian Fluid Mechanics. The basic molecular models have been discussed in the chapter on Polymer Physics: the Rouse and Zimm models [4]. Here, we present some simple physical interpretations of the polymer rheological behavior.

8.2.1 Polymer Solution as a Suspension

The simplest explanation of the polymeric liquid viscosity is that of a dilute solution. A dilute solution is like a suspension of colloidal particles, except that the particles are not spherical. The polymers in equilibrium have a coil-like structure (see chapter on Polymer Physics). Considering the polymer coils to represent colloidal particles, the viscosity can be related to the volume fraction. For dilute colloidal suspension, the viscosity of the suspension is more than the solvent viscosity, and is given by the Einstein expression

$$\eta = \eta_s \left(1 + \frac{5}{2} \phi \right), \tag{8.21}$$

where ϕ is the volume fraction of the colloidal particle. This approximation is valid in the limit $\phi \to 0$ for spherical particles. Polymer coils are not rigid spheres, but behave more like a porous sphere (on the average, because the shape of the coil changes continuously due to thermal Brownian forces from solvent). Because of this the increase in viscosity is reduced. For dilute polymers, the viscosity can be written as

$$\eta = \eta_s \left(1 + U_{\eta R} \phi \right), \tag{8.22}$$

where $U_{\eta R}$ is a constant $<5/2$. This constant is a universal constant for all polymers and solvents chemistry and depends only on the solvent quality (which could be poor, theta, or good solvents). The Zimm theory predicts this constant to be 1.66, whereas experiments and simulations incorporating fluctuating hydrodynamic interactions (without the approximations made in Zimm theory) observe the value to be $U_{\eta R} = 1.5$.

8.2.2 Shear Thinning

As noted before, the extent of shear thinning is much more significant in concentrated solutions and melts in comparison with dilute solutions (see Fig. 8.7). The explanation for viscosity decrease in dilute solutions is subtle and involves several factors [4, 7]. We will restrict to simple explanations in entangled systems (i.e. in concentrated solutions and melts). In the entangled state, the microstructure of the solution is like a network: the entanglements act like nodes of a covalently bonded network junction which restricts the motion of the polymers and therefore of the solution. This is the reason for the high viscosity of the solution. However, upon shearing, the chains begin to de-entangle and align along the flow. This reduces the viscosity.

8.2.3 Normal Stresses

The existence of normal stresses in polymeric systems is due to the anisotropy induced in the microstructure because of flow. This can be easily understood in the case of dilute polymer solutions. In the absence of flow, the coil-like structure of the molecules assumes a spherical pervaded volume on the average. The flow causes the molecules to stretch towards the direction of flow and tumble. This results in a pervaded volume that is ellipsoidal and oriented towards the direction of flow. The restoring force is different in the two planes xx and yy. This results in the anisotropic normal forces.

8.2.4 Elongational Viscosity

As shown in Fig. 8.11, dilute solutions have a dramatic effect on the elongational viscosity. The main reason for this is that in the dilute state at equilibrium, the polymers are in a coil like state in a suspension whose viscosity is only slightly more than the solvent viscosity, whereas upon elongation, the chains are oriented in the direction of elongation. After a critical elongation rate given by the Weissenberg number Wi $= 0.5$, the thermal fluctuations can no longer hold against the forces generated on the ends of the polymer due to flow. The chains then begin to stretch. This leads to anisotropy and stress differences between the elongation and compres-

Fig. 8.12 Theoretical
prediction of the elongational
viscosity in the "standard
molecular theory" or the tube
model, for linear polymer
melts

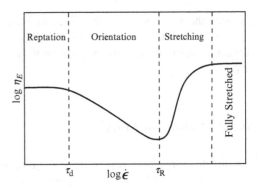

sion planes, and is observed as an increase in the elongational viscosity in (8.18).
Since the polymers are not infinitely extensible, the stress difference between the
planes saturates to the value given by that in the fully stretched state (which is like
a slender rod).

The behavior of elongational viscosity in concentrated solutions and melts is different. The zero-shear rate viscosity is itself much higher than the solvent viscosity.
A semi-quantitative model for the behavior in linear polymer melts can be interpreted in terms of the tube model or also called as the "standard molecular theory"
[6]. At small strain rates compared to the inverse reptation time (or disentanglement
time) τ_d^{-1}, thermal fluctuations are stronger than the hydrodynamic drag forces induced by the flow, and the chains remain close to equilibrium microstructure. The
viscosity is nearly same as the zero-shear rate viscosity, as shown schematically
in Fig. 8.12. At higher strain rates $\dot{\varepsilon} \ll \tau_d^{-1}$, the tubes (the pervaded volumes of
the polymers) are disentangled and begin to orient along the extensional direction,
which decreases the viscosity. This continues till the strain rate becomes comparable to the inverse Rouse relaxation time τ_R^{-1}, when chain stretching begins (similar
to the case in dilute polymer solutions). This causes an increase in the viscosity.
The terminal region of constant viscosity occurs due to the chains attaining their
maximum stretch. Most of these theoretical predictions have not been verified by
experiments, mainly due to the difficulty in consistently measuring the elongational
viscosity at very large strain rates and strains.

Other Reading

An historical account of rheology, in general, is discussed in Ref. [11]. Ref. [7]
contains an extensive literature survey of various aspects of polymer rheology
and general rheology. A compilation of recent works in melt rheology is given in
Ref. [9].

Figures in this chapter

The figures in this chapter have been schematically redrawn for educational purposes. They may not bear direct comparisons with absolute values in reality. Where possible, references to the sources containing details or other original references have been provided.

References

1. Barnes HA (2000) Handbook of elementary rheology. University of Wales, Inst of Non-Newt Fluid Mech, Aberystwyth
2. Barnes HA, Hutton JF, Walters K (1989) An introduction to rheology. Elsevier, Amsterdam
3. Bird RB, Armstrong RC, Hassager O (1987) Dynamics of polymeric liquids – Vol 1: Fluid mech, 2nd edn. Wiley, New York
4. Bird RB, Curtiss CF, Armstrong RC et al (1987) Dynamics of polymeric liquids – Vol 2: Kinetic theory, 2nd edn. Wiley, New York
5. Dealy JM (1995) Official nomenclature for material functions describing the response of a viscoelastic fluid to various shearing and extensional deformations. J Rheol 39:253–265
6. Dealy JM, Larson RG (2006) Structure and rheology of molten polymers: From structure to flow behaviour and back again. Hanser Publishers, Munich
7. Larson RG (1999) The structure and rheology of complex fluids. Oxford University Press, New York
8. Macosko CW (1994) Rheology: Principles, measurements, and applications. Wiley, NY
9. Piau J-M, Agassant J-F (1996) Rheology for polymer melt processing. Elsevier, Amsterdam
10. Rubinstein M, Colby RH (2003) Polymer physics. Oxford University Press, New York
11. Tanner RI, Walters K (1998) Rheology: An historical perspective. Elsevier, Amsterdam

Chapter 9
Active Matter

Gautam I. Menon

Abstract The term *active matter* describes diverse systems, spanning macroscopic (e.g., shoals of fish and flocks of birds) to microscopic scales (e.g., migrating cells, motile bacteria and gels formed through the interaction of nanoscale molecular motors with cytoskeletal filaments within cells). Such systems are often idealizable in terms of collections of individual units, referred to as active particles or self-propelled particles, which take energy from an internal replenishable energy depot or ambient medium and transduce it into useful work performed on the environment, in addition to dissipating a fraction of this energy into heat. These individual units may interact both directly and through disturbances propagated via the medium in which they are immersed. Active particles can exhibit remarkable collective behavior as a consequence of these interactions, including non-equilibrium phase transitions between novel dynamical phases, large fluctuations violating expectations from the central limit theorem and substantial robustness against the disordering effects of thermal fluctuations. In this chapter, following a brief summary of experimental systems, which may be classified as examples of active matter, I describe some of the principles which underlie the modelling of such systems.

9.1 Introduction to Active Fluids

Anyone who has admired the intricate dynamics of a group of birds in flight or the co-ordinated, almost balletic manoeuvres of a school of swimming fish can appreciate the motivation for the study of "active matter": How do individual self-driven units, such as wildebeest, starlings, fish or bacteria, flock together, generating large-scale, spatiotemporally complex dynamical patterns? [51] What are the rules which govern this dynamics and how do the principles of physics constrain the behavior

G.I. Menon (✉)
The Institute of Mathematical Sciences, CIT Campus, Taramani, Chennai 600113, India
e-mail: menon@imsc.res.in

J.M. Krishnan et al. (eds.), *Rheology of Complex Fluids*,
DOI 10.1007/978-1-4419-6494-6_9, © Springer Science+Business Media, LLC 2010

of each such unit? Finally, what are the simplest possible models for such behavior and is there any commonality to the description of these varied problems? [46–48]

Describing these diverse problems in terms of individual "agents", which evolve via a basic set of update rules while interacting with other agents, provides a general way of approaching a large number of unrelated problems. These include the description of the propagation of infectious diseases in a population, the seasonal migration of animal populations, the collective motion and co-ordinated activities of groups of ants and bees and the swimming of shoals of fish [48]. Of these problems, the subset of problems involving agents whose *mechanical* behavior at a scale larger than the individual agent must be constrained by local conservation laws, such as the conservation of momentum, forms a special class [44]. It is these systems that are the primary focus of this chapter.

Generalizing from the examples above, a tentative definition of active matter might be the following: *Active matter is a term which describes a material (either in the continuum or naturally decomposable into discrete units), which is driven out of equilibrium through the transduction of energy derived from an internal energy depot or ambient medium into work performed on the environment.* Such systems are generically capable of emergent behavior at large scales. The hydrodynamic description of such problems should be equally applicable to situations where the granularity of the constituent units is not resolved, but which are rendered non-equilibrium in a qualitatively similar way [39, 48].

What differentiates active systems from other classes of driven systems (sheared fluids, sedimenting colloids, driven vortex lattices, etc.) is that the energy input is internal to the medium (i.e., located on each unit) and does not act at the boundaries or via external fields. Further, the direction in which each unit moves is dictated by the state of the particle and not by the direction imposed by an external field.

Such individual units are, in general, anisotropic, as in Fig. 9.1. Collections of such units are thus capable of exhibiting orientationally ordered states. A canonical example of an ordered state of orientable units obtained in thermal equilibrium is the nematic liquid crystal, in which anisotropic particles align along a common axis. Active systems of such units can, in addition, exhibit directed motion along this axis; a concrete example is illustrated in Fig. 9.1. In the context of active particles in a fluid, the terminology "swimmers" or "self-propelled particles" is often used,

Fig. 9.1 A school of fish, illustrating the tendency towards parallel alignment while swimming. Picture courtesy Prof. R. Kent Wenger

while the terms "active nematic" or "living liquid crystals" occur in the discussion of the orientationally ordered collective states of active particles [39].

Why study active systems? For one, such systems can display phases and phase transitions absent in systems in thermal equilibrium [46, 47]. For another, active matter often exhibits unusual mechanical properties, including strong instabilities of ordered states to small fluctuations. Such instabilities may be *generic* in the sense that they should appear in any hydrodynamic theory, which enforces momentum conservation and includes the lowest order contribution to the system stress tensor arising from activity [44]. Fluctuations in active systems are generically large, often deviating qualitatively from the predictions of simple arguments based on the central limit theorem [37, 45–47]. In common with a large number of related non-equilibrium systems, some examples of active matter appear to be self-tuned to the vicinity of a phase transition, where response can be anomalously large and fluctuations dominate the average behavior.

The study of active matter spans many scales. The smallest scales involve the modelling of individual motile organisms, such as fish or individual bacteria or even nanometer-scale motor proteins, such as kinesin or dyneins which move in a directed manner along cytoskeletal filaments. Hydrodynamic descriptions of large numbers of motile organisms operate at a larger length scale, averaged over a number of such swimmers that is large enough for the granularity at the level of the individual swimmer to be neglected, while still allowing for the possibility of spatial fluctuations at an intermediate scale.

Finally, all *living* matter is matter out of thermodynamic equilibrium. The source of this non-equilibrium is, typically, the hydrolysis of an nucleoside tri-phosphate (NTP, such as ATP or GTP) molecule into its di-phosphate form, releasing energy. This energy release can drive conformational changes in a protein, leading to mechanical work. Biological matter is thus generically internally driven, precisely as demanded by our definition of active matter. Understanding the general principles which govern active matter systems could thus provide ideas, terminology and a consistent set of methods for the modelling of the mechanical behavior of living, as opposed to dead, biological matter. This is perhaps the most substantial motivation for the study of active matter.

The outline of this chapter is the following: Section 9.2 provides some examples of matter which can be classified as active, using our definition. Section 9.3 summarizes results on a simple model for an individual swimmer and summarizes necessary ingredients for a coarse-grained description of a large number of swimmers. Section 9.4 begins by illustrating the derivation of the equations of motion and the stress tensor of a simple fluid and then goes on to illustrate how this derivation may be extended to describe fluids with orientational order close to thermal equilibrium. It then goes on to discuss how novel contributions to the fluid stress tensor can arise from active fluctuations, deriving an equation of motion for small fluctuations above a pre-assumed nematic state. These calculations lead to predictions for the rheological behavior of active systems and the demonstration of the instability of the ordered nematic (or polar) state to small fluctuations. In Sect. 9.5, the theory of active gels is summarized briefly and some results from this formalism are outlined. Section 9.6 provides a brief summary.

9.2   Active Matter Systems: Some Examples

The subsections below list a (non-exhaustive) set of examples of active matter systems. The large-scale behavior of some of these systems is the focus of later sections.

9.2.1   Dynamical Behavior in Bacterial Suspensions

Dombrowski et al. study the velocity fields induced by bacterial motion at the bottom of sessile and pendant drops containing the bacterium *B. subtilis* [12]. These flows reflect the interplay between bacterial chemotaxis and buoyancy effects which together act to carry bioconvective plumes down a slanted meniscus, concentrating cells at the drop edge (sessile) or at the drop bottom (pendant). The motion exhibited by groups of bacteria as a result of this self-concentration is large-scale at the level of individual bacteria, exhibiting vortical structures and other complex patterns as a consequence of the hydrodynamic interaction between swimming bacteria. Thus, these experiments highlight the crucial role of hydrodynamics, coupled with the motion of individual active swimmers, in generating self-organized, large-scale dynamical fluctuations in the surrounding fluid.

9.2.2   Mixtures of Cytoskeletal Filaments and Molecular Motors

The influential experiments of Nedelec and collaborators combine a limited number of cytoskeletal and motor elements – microtubules, kinesin complexes and ATP – in a two-dimensional geometry in vitro finding remarkable self-organized patterns evolving from initially disordered configurations [32]. These include individual asters and vortices as well as disordered arrangements of such structures, in addition to bundles and disordered states at varying values of motor densities. These are self-organized structures, large on the scale of the individual cytoskeletal filament and motor, which require ATP (and thus are non-equilibrium) in order to form and be sustained. Several hydrodynamical approaches to this problem have been proposed, as in [2,23,41]. These experiments and associated theory are summarized in [18].

9.2.3   Layers of Vibrated Granular Rods

The experiments of Narayanan et al. take elongated rod-like copper particles in a quasi-two-dimensional geometry, and vibrate them vertically in a shaker [31]. This agitated monolayer of particles is maintained out of equilibrium by the shaking, which effectively acts to convert vertical motion into horizontal motion via the tilting of the rod. The particles are back–front symmetric, and are thus nematic in character. These experiments see large, dynamical regions which appear to fluctuate coherently. Similar behavior is predicted in theories of flocking behavior induced purely by increasing the concentration in dense aggregates of particles held out of equilibrium [48, 51]. These experiments provide a particularly striking signature of non-equilibrium steady state behavior in "flocking" systems: the presence of number fluctuations in such systems which scale anomalously with the size of the region being averaged, i.e., with a power of N, which exceeds the central limit prediction of a $\sqrt{N}$ behavior of number fluctuations [48].

9.2.4   Fish Schools

Experiments of Makris and collaborators use remote sensing methods on a continental scale to access the structure and dynamics of large-scale shoals of fish, containing an order of a few million individuals [25]. Apart from the relatively rapid time-scale for the reorganization of the shoal – around 1–10 min in their experiments – these experiments also see evidence for "fish waves"; propagating internal disturbances within the shoal which occur at relatively regular intervals, representing disturbances at scales far larger than that of the individual fish. The speeds of such waves are larger, by around a factor of 10, than the velocities of the swimming fish. They appear to represent "locally interconnected compaction events" similar to the Mexican waves exhibited through the co-ordinated motion of spectators in stadia. It is interesting that similar wave-like excitations are predicted in the hydrodynamic theories of [44, 47], where they involve waves of concentration and splay [47] or splay–concentration and bend [44].

9.2.5   Bird Flocks

The STARFLAG collaboration has imaged large flocks of starlings (between 10^3 and 10^4 individuals at a time), using computer-aided imaging techniques to understand the dynamics of individual birds and how this dynamics is influenced by the spatial distribution and behavior of neighboring birds. Surprisingly, and contrary to what might have been naively expected, birds appear to adjust to the motion of the flock by measuring the behavior of topologically (and not metrically) related neighbors [3]. Thus, at each instant, each bird appears to be comparing its instantaneous position and velocity to those of the 5–7 birds closest to it, making the adjustments required to maintain the coherence of the flock. This strategy has the advantage that reducing the density of the flock should not then impact the coherence of the flock, since only topological and not metrical relationships are involved, a fact that calls into question the gradient expansions favored by most theoretical work, which represents flocking behavior by coarse-grained equations of motion for a few hydrodynamic fields.

9.2.6 Marching Behavior of Ants and Locusts

Ian Couzin's group at Princeton has investigated the transition to marching behavior in locusts. Swarms of the desert locust, *Schistocerca gregaria*, in their non-flying form, can exist in a relatively solitary individualistic state as well as a gregarious collective state. In the "gregarious" state, such locusts can form huge collective marching units, which forage all vegetation in their path. This has a huge social and economic impact on humans, affecting the livelihood of one in ten people on the planet in plague years. These experiments provide an experimental example of the collective transition in simple computational models of flocking behavior. From more recent work from this group, it appears, somewhat unusually, that the transition is induced by "cannibalistic" behavior, in which a locust is successful in biting the rear-quarters of the locust immediately in front [7, 10].

9.2.7 Listeria monocytogenes Motility

The bacterium *Listeria monocytogenes* is a simple model system for cell motility, which derives its ability to move from the polymerization of actin, leading to the formation of a "comet tail" emerging from the rear of the bacterium [35]. Motility appears to arise from the deformation of the gel formed by the actin and cross-linking proteins as a consequence of continuing polymerization, which results in a propulsive force on the bacterium. The non-equilibrium comes from ATP-driven actin polymerization. Interestingly, many features of the experiment can be reproduced in in vitro systems where actin polymerization is initiated at the surface of specially treated beads, which then exhibit symmetry-breaking motility [29].

9.2.8 Cell Crawling

There is a vast and intriguing literature on cell crawling on substrates [8, 9, 14, 50]. Such crawling appears to have four basic steps: the extension of cellular protrusions, the attachment to the substrate at the leading edge, the translocation of the cell body and the detachment at the rear. Cell crawling is largely mediated by a meshwork of actin in gel form, whose fluidity is actively maintained through the action of myosin motors and other associated proteins [8, 50]. Interestingly, motility has also been observed in nucleus-lacking cell fragments excised by lasers over a period of several hours.

9.2.9 Active Membranes

Experiments on fluctuating giant vesicles containing bacteriorhodopsin (BR) pumps reconstituted in a lipid bilayer indicate that the light-driven proton pumping activity of BR amplifies membrane shape fluctuations [26, 27]. The BR pumps transfer protons in one direction across the membrane as they change conformation upon excitation by light of a specific wavelength. These experiments have been described in terms of a non-equilibrium "active" temperature [26, 27]. Hydrodynamic theories which interpret the experiments *via* a description of a membrane with a density of embedded diffusing dipolar force centres, *i.e.* as an "active membrane", have also been studied in some detail. It is interesting that several developments in the theoretical description of generic active matter reflect ideas first introduced in the context of active membranes [33, 38, 42]. A detailed study of a biologically relevant model of an active membrane system which contains a large number of references to the literature is available in [21].

9.3 The Swimmer: Individual and Collective

The examples provided in the previous section are indicative of some of the diversity exhibited by systems which fall under the general category of active matter. These can broadly be classified into systems in which the role of mechanical conservation laws (e.g., for momentum) is important and systems in which, typically, only the conservation of number is relevant.

The flocking model of Vicsek and collaborators [51], recast in terms of hydrodynamic equations of motion by Toner and Tu [46–48], is an agent-based model in which cooperativity is driven by a density-dependent interaction, which tends to orient individual units in the direction in which their neighbors move. The environment — the ground and vegetation for the moving locusts, for example — provides merely the background in this case and has no other dynamical significance. In the case of the flock of starlings, the communication between individual starlings does not appear to be primarily via the medium, but through visual contact.

The case of the motile bacterium and the swimming fish, on the other hand, are cases where the medium plays an important role in transferring momentum to and from the swimmers and in determining the interactions between swimmers. This case will be discussed below, first for the situation of the individual swimmer and then for collections of swimmers, modeled via a hydrodynamic approach.

9.3.1 The Individual Swimmer

The central idea in the modelling of individual swimmers is that the system must be force-free, when averaged over lengthscales larger than the characteristic dimension of the swimmer [6, 39]. This is a consequence of Newton's third law, which imposes that the force exerted by the swimmer on the fluid must be equal and opposite to the force exerted by the fluid on the swimmer. Thus, if one considers the swimmer as a source of forces locally within the fluid which act upon the fluid, such a source cannot have a monopole component but may have a dipole (or higher multipole) component.

Depending on the character of the swimmer, more distinctions are possible [6, 39]. Contractile swimmers or pullers (such as bacteria propelled by flagella at the head of the organism) pull fluid in along the long axis and push it out along an axis normal to their midpoint. Tensile swimmers or pushers push fluid out along their long axis and pull fluid in along the midpoints. They are propelled from the rear, justifying the terminology of pushers.

Consider a simple model for a microscopic swimmer. In the Stokes limit, time does not enter these equations explicitly and the velocity field is completely specified by the boundary conditions imposed on the flow. (The use of the Stokes limit is generically justified in the case of bacteria, where characteristic Reynolds numbers at the scale of a single particle are of the order of 10^{-4} or smaller.) Reversing the velocity field at the boundaries should retrace the velocity field configuration, implying that the trajectory assumed by the swimmer in its configuration space cannot be time-reversal invariant. The helical or "corkscrew-like" motion of the flagellae of the bacterium *E. coli* (see Fig. 9.2), discussed by Purcell [34], provides

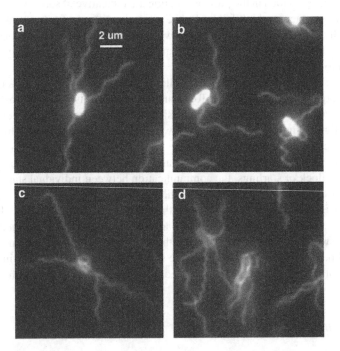

Fig. 9.2 Cells of *Escherichia coli* labelled with a dye and examined in a fluorescence microscope. Each of the sub-figures, labelled (**a**), (**b**), (**c**) and (**d**) refers to a different set of cells. From these images, the behavior of the flagellae of individual cells can be studied as *E. coli* swims [49]. Picture courtesy Prof. Howard Berg

a particularly attractive example of how the limitations on directed motion at the
low Reynolds numbers required for the Stokes limit approximated to be valid can
be overcome.

Averaging over multiple strokes of the swimmer simplifies the description: the
broken temporal symmetry required for translation in a flow governed by the Stokes
equations can be replaced by a broken spatial symmetry. This singles out the direction of motion of the swimmer. Following a model introduced recently by Baskaran
and Marchetti (whose notations and treatment we follow closely in this section), the
swimmer is modelled as an asymmetric, rigid dumbbell [6]. This dumbbell consists
of two differently sized spheres, of radii a_L (large) and a_S (small), forming the head
and the tail of the swimmer. The length ℓ of the swimmer is the length between the
two centres. The orientation of the swimmer is given by the unit vector $\hat{v}$, drawn
from the smaller sphere to the larger sphere.

Thus, the equations of motion of the two spheres, with locations $\mathbf{r}_{L\alpha}$ and $\mathbf{r}_{R\alpha}$, are
given by

$$\partial_t \mathbf{r}_{L\alpha} = \mathbf{u}(\mathbf{r}_{L\alpha})$$
$$\partial_t \mathbf{r}_{S\alpha} = \mathbf{u}(\mathbf{r}_{S\alpha}), \tag{9.1}$$

where the constant length between the two spheres is imposed by the constraint that
$\mathbf{r}_{L\alpha} - \mathbf{r}_{S\alpha} = \ell \hat{v}_\alpha$ and the velocity field is given by $\mathbf{u}(\mathbf{r})$. The no-slip condition at the
surfaces of the spheres requires that the sphere move with the velocity of the fluid
next to it.

The velocity field obeys the Stokes equation and is constrained by
incompressibility,

$$\eta \nabla^2 \mathbf{u}(\mathbf{r}) = \nabla p + \mathbf{F}_{active} + \mathbf{f}_{Noise},$$
$$\nabla \cdot \mathbf{u}(\mathbf{r}) = 0, \tag{9.2}$$

where the "force" terms which enter on the right-hand side are composed of a term,
which is associated purely with activity (i.e., vanishes in thermal equilibrium) as
well as of a second "noise" term modelling the fluctuations in fluid velocity arising from purely thermal fluctuations. Such noise terms do not conventionally enter
the Navier–Stokes equations, but must generically be included in a coarse-grained
description. Detailed expressions for these forces are available in [6].

To solve the Stokes equations, we insert a delta function force $\delta(\mathbf{r})\mathbf{F}$ on the right-
hand side, representing a fundamental source term from which more complex force
configurations can be constructed. In Fourier space, the incompressibility condition
is imposed as $\mathbf{q} \cdot \mathbf{u}(\mathbf{q}) = \mathbf{0}$, while the Stokes equations are $-\eta q^2 \mathbf{u}(\mathbf{q}) + i\mathbf{q}p(\mathbf{q}) =$
$\mathbf{F}(\mathbf{q})$. This then gives

$$\mathbf{u}(\mathbf{r}) = \frac{1}{\eta q^2}\left(\mathbf{F} - \frac{(\mathbf{q} \cdot \mathbf{F})\mathbf{q}}{q^2}\right), \tag{9.3}$$

with

$$p(\mathbf{q}) = i\frac{\mathbf{q} \cdot \mathbf{F}}{q^2}. \tag{9.4}$$

These are inverted by

$$\mathbf{u}(\mathbf{r}) = \frac{1}{8\pi\eta}\left(\frac{\boldsymbol{\delta}}{r} + \frac{\mathbf{rr}}{r^3}\right) \cdot \mathbf{F}. \tag{9.5}$$

The $\boldsymbol{\delta}$ symbol is the unit tensor. In a Cartesian basis of $\mathbf{i}, \mathbf{j}, \mathbf{k}$ it is

$$\boldsymbol{\delta} = \mathbf{ii} + \mathbf{jj} + \mathbf{kk}. \tag{9.6}$$

The quantity acting on the force on the right-hand side is the Oseen tensor, defined through

$$\mathscr{O}_{ij}(\mathbf{r}) = \left(\delta_{ij} + \hat{r}_i\hat{r}_j\right)/8\pi\eta r \tag{9.7}$$

for $|r| > a_{\mathrm{L,S}}$, with $\hat{r} = \mathbf{r}/|\mathbf{r}|$ a unit vector. The Stokes equations are solved by the superposition

$$u_i(\mathbf{r}) = f\sum_\alpha \left[\mathscr{O}_{ij}(\mathbf{r} - \mathbf{r}_{\mathrm{L}\alpha}) - \mathscr{O}_{ij}(\mathbf{r} - \mathbf{r}_{\mathrm{S}\alpha})\right]\hat{v}_{\alpha j}. \tag{9.8}$$

The divergence at short distances is eliminated through the definition: $\mathscr{O}_{ij}(|\mathbf{r}| \le a_{\mathrm{L,S}}) = \delta_{ij}/\zeta_{\mathrm{L,S}}$, where $\zeta_{\mathrm{L,S}} = 6\pi\eta a_{\mathrm{L,S}}$.

The dynamics of an *extended* body in Stokes flow follows from translation of the hydrodynamic centre and rotations about the hydrodynamic centre. (The hydrodynamic centre refers to the point about which the net hydrodynamic torque on the body vanishes; it thus plays the same role in Stokes flow as the centre of mass in inertial dynamics.) For the problem of a rigid dumbbell in an external flow, the hydrodynamic centre is obtained from

$$\mathbf{r}^{\mathrm{C}} = \frac{\zeta_{\mathrm{L}}\mathbf{r}_{\mathrm{L}} + \zeta_{\mathrm{S}}\mathbf{r}_{\mathrm{S}}}{\zeta_{\mathrm{L}} + \zeta_{\mathrm{S}}} = \frac{a_{\mathrm{L}}\mathbf{r}_{\mathrm{L}} + a_{\mathrm{S}}\mathbf{r}_{\mathrm{S}}}{a_{\mathrm{L}} + a_{\mathrm{S}}}. \tag{9.9}$$

The equations of motion for the translation and rotation of the hydrodynamic centre follow from

$$\partial_t \mathbf{r}_\alpha^{\mathrm{C}} = v_0\hat{v}_\alpha + \frac{1}{\zeta}\sum_{\beta\ne\alpha}\mathbf{F}_{\alpha\beta} + \Gamma_\alpha(t)$$

$$\omega_\alpha = \frac{1}{\zeta_{\mathrm{R}}}\sum_{\beta\ne\alpha}\tau_{\alpha\beta} + \Gamma_\alpha^{\mathrm{R}}(t), \tag{9.10}$$

where the angular velocity describing rotations about the hydrodynamic centre is defined by

$$\partial_t \hat{v}_\alpha = \hat{v}_\alpha \times \omega_\alpha \tag{9.11}$$

and

$$\bar{\zeta} = (\zeta_L + \zeta_S)/2, \zeta_R = \ell^2 \bar{\zeta}. \tag{9.12}$$

The random forces Γ_α and Γ_α^R lead to diffusion at large length scales [6].

The forces $F_{\alpha\beta}$ and the torques $\tau_{\alpha\beta}$ arise from hydrodynamic couplings between swimmers. An isolated swimmer is propelled at speed

$$v_0 = -\frac{f \Delta a}{8\pi \eta \ell \bar{a}} \tag{9.13}$$

with velocity $v_0 \hat{v}$. This velocity arises purely as a consequence of the fact that the hydrodynamic and the geometric centres do not coincide. For symmetric swimmers, this velocity is zero and the swimmer is a "shaker" as opposed to a "mover."

The interactions between swimmers can be calculated in the dilute limit by a multipole expansion, yielding

$$F_{12} \approx 2f\bar{a}\ell \left[3(\hat{r}_{12} \cdot \hat{v}_2)^2 - 1\right] \frac{\hat{\mathbf{r}}_{12}}{r_{12}^2} \tag{9.14}$$

for the hydrodynamic force exerted by the βth swimmer on the αth one. In addition, expressions for the hydrodynamic torque between swimmers in the dilute limit can be derived and are presented in [6].

The hydrodynamic force decays as $1/r_{12}^2$, as follows from its dipole character. The torque consists of two terms: One is nonzero even for shakers and aligns swimmers regardless of their polarity. The second term vanishes for symmetric swimmers, serving to align swimmers of the same polarity. A more detailed discussion of the structure of the forces and torques for pushers and pullers is available in [6].

9.3.2 Multiple Swimmers

The coarse-grained version of the many swimmer problem is defined through the following local fields. First, we must have a density of active particles, defined microscopically in terms of

$$c(\mathbf{r}, t) = \left\langle \sum_\alpha \delta(\mathbf{r} - \mathbf{r}_\alpha^C(t)) \right\rangle. \tag{9.15}$$

Second, in the case in which the axes of a fore–aft asymmetric swimmer are largely aligned along a common direction – as in a magnet – we can define a local field describing polar order in the following way:

$$\mathbf{P}(\mathbf{r}, t) = \frac{1}{c(\mathbf{r}, t)} \left\langle \sum_\alpha \hat{v}_\alpha \delta(\mathbf{r} - \mathbf{r}_\alpha^C(t)) \right\rangle. \tag{9.16}$$

Third, and finally, when one considers ensembles of interacting swimmers, we must also consider the possibility of additional and more subtle *macroscopic* variables representing orientational order. An ensemble of individual particles, each aligned, on average, along a common *axis*, is familiar in soft condensed matter physics. Such systems are referred to as nematics and the ordering as nematic ordering. In a *nematic* order, the alignment is along a common axis but a vectorial direction is not picked out.

Orientational order in the nematic phase is generally described by a second-rank, symmetric traceless tensor $Q_{\alpha\beta}(\mathbf{x}, t)$, defined in terms of the second moment of the microscopic orientational distribution function. This (order-parameter) tensor can be expanded as

$$Q_{\alpha\beta} = \frac{3}{2} S \left(n_\alpha n_\beta - \frac{1}{3}\delta_{\alpha\beta} \right) + \frac{1}{2} T \left(l_\alpha l_\beta - m_\alpha m_\beta \right). \qquad (9.17)$$

The three principal axes of this tensor, obtained by diagonalizing $Q_{\alpha\beta}$ in a local frame, specify the direction of nematic ordering $\mathbf{n}$, the codirector $\mathbf{l}$ and the joint normal to these, labelled by $\mathbf{m}$. The principal values S and T represent the strength of ordering in the direction of $\mathbf{n}$ and $\mathbf{m}$, quantifying, respectively, the degree of uniaxial and biaxial nematic orders.

In thermal equilibrium, the energetics of $Q_{\alpha\beta}$ is calculated from a Ginzburg–Landau functional, first proposed by de Gennes, based on an expansion in rotationally invariant combinations of $Q_{\alpha\beta}$ and its gradients [11]. The Ginzburg–Landau–de Gennes functional F is

$$F = \int d^3\mathbf{x} \left[\frac{1}{2} A Tr\mathbf{Q}^2 + \frac{1}{3} B Tr\mathbf{Q}^3 + \frac{1}{4} C(Tr\mathbf{Q}^2)^2 \right.$$
$$\left. + E'(Tr\mathbf{Q}^3)^2 + \frac{1}{2} L_1 (\partial_\alpha Q_{\beta\gamma})(\partial_\alpha Q_{\beta\gamma}) \right]. \qquad (9.18)$$

Here, $A = A_0(1 - T/T^*) \, T^*$ denoting the supercooling transition temperature, A_0 is a constant, L_1 is an elastic constant and α, β, γ denote the Cartesian directions. Other elastic terms can also be included; this simple approximation corresponds to what is called the one Frank constant approximation.

Two simplifications are possible and often convenient. First, we may assume-uniaxial rather than biaxial order, since this is by far the more common form of ordering. In the ordered nematic state, the average orientation occurs along a direction $\hat{n}$. This is the nematic director, defined to be a unit vector. We can thus work within a description in which the components of $Q_{\alpha\beta}$ are written out in terms of the components of the nematic director $\mathbf{n}$. This gives us

$$Q_{\alpha\beta} = \frac{3S}{2} \left(n_i n_j - \frac{1}{3}\delta_{ij} \right). \qquad (9.19)$$

The energetics of small deviations from the aligned state is obtained, in this representation, from a Frank free energy appropriate to uniaxial nematics:

$$f_{FO} = \frac{1}{2}K_1(\text{div } \mathbf{n})^2 + \frac{1}{2}K_2(\mathbf{n} \cdot \text{curl } \mathbf{n})^2 + \frac{1}{2}K_3(\mathbf{n} \times \text{curl } \mathbf{n})^2. \qquad (9.20)$$

Here, K_1 is the splay elastic modulus, associated with a splay deformation $\nabla \cdot \mathbf{n}$, K_2 is the twist elastic modulus and K_3 is the bend elastic modulus. The Frank constants K_1, K_2 and K_3 have the dimensions of a force and can be represented as the ratio of an energy to a length scale. We can assume $K_i \sim k_B T_c / a$ where a is a molecular length of order 1 nm. Note that this description uses three elastic constants (which can be reduced to the single elastic coefficient of (9.18) by assuming that $K_1 = K_2 = K_3$).

With this background, the definition of a local field representing nematic order follows from [11],

$$Q_{ij} = \frac{1}{c(\mathbf{r}, t)} \left\langle \sum_\alpha \left(\hat{v}_{\alpha i} \hat{v}_{\alpha j} - \frac{1}{3}\delta_{ij} \right) \delta(\mathbf{r} - \mathbf{r}_\alpha^C(t)) \right\rangle. \qquad (9.21)$$

9.4 Hydrodynamic Approaches to Active Matter

A class of questions relating to the modelling of active matter is concerned with (a) whether forms of collective ordering are at all possible in ensembles of interacting active particles and (b) whether such ordering, if assumed to preexist, can be shown to be stable against fluctuations. Our definition of the single swimmer associated a direction with the swimming motion, the direction of the axis formed by connecting, say, the centre of the small sphere to the centre of the large sphere. The question is thus whether the axes or orientations of different swimmers can be aligned as a consequence of their interaction.

In subsections below, the problem of deriving equations of motion for the nematic order parameter field and the construction of a stress tensor appropriate to a nematic fluid are briefly examined. Results for the active nematic are summarized and the significance of these results for the rheological properties of active matter is briefly outlined.

9.4.1 Equations of Motion: Fluid and Nematic

Identifying conservation laws and broken symmetries is the crucial first step in constructing hydrodynamic equations of motion for the relevant fields in the problem. For both conserved and hydrodynamic fields, the relaxation of long wavelength fluctuations proceeds slowly, with the relevant timescales for the relaxation of the fluctuation diverging as the wavelength of the perturbation approaches infinity.

For a simple fluid with no internal structure, the conservation laws for the local energy ε, the density ρ and the three components of the momentum density $\mathbf{g}_i$ are

$$\frac{\partial \varepsilon}{\partial t} = -\nabla \cdot j^\varepsilon,$$

$$\frac{\partial \rho}{\partial t} = -\nabla \cdot g,$$

$$\frac{\partial \mathbf{g}_i}{\partial t} = -\nabla_j \pi_{ij}, \tag{9.22}$$

where j^ε is the energy current and π_{ij} is the momentum current tensor, sometimes also called the stress tensor. The conserved momentum density itself acts as a current for another conserved density, the mass (equivalently, number) density, a relation which is responsible for sound waves in fluids.

The hydrodynamic description of fluids with internal order (such as the nematic or polar fluid) must account for additional hydrodynamic modes arising out of the fact that the ordering represents a broken symmetry. For small deviations from equilibrium, one derives an equation for entropy generation and casts it in terms of the product of a flux and a force. Such fluxes must vanish at thermodynamic equilibrium. Close to equilibrium, it is reasonable to expect that fluxes should have a smooth expansion in terms of forces.

As an illustrative example, consider the simple fluid in the absence of dissipation. We have, with $\mathbf{u}$ the velocity field,

$$\mathbf{g} = \rho \mathbf{u},$$

$$\pi_{ij} = p\delta_{ij} + u_j g_i$$

$$j^\varepsilon = (\varepsilon + p)\mathbf{u} = \left(\varepsilon_0 + p + \frac{\rho u^2}{2} \right). \tag{9.23}$$

The mass conservation equation is just

$$\frac{\partial \rho}{\partial t} = -\nabla \cdot (\rho \mathbf{u}), \tag{9.24}$$

while the momentum conservation equation is

$$\frac{\partial \mathbf{g}_i}{\partial t} = \frac{\partial \rho u_i}{\partial t} = -\nabla_j \pi_{ij} = -\nabla_i p - \nabla_j (\rho u_i u_j). \tag{9.25}$$

This is Euler's equation, usually written as

$$\frac{\partial \mathbf{u}}{\partial t} + (\mathbf{u} \cdot \nabla)\mathbf{u} = \frac{-1}{\rho} \nabla p. \tag{9.26}$$

The dissipative contribution to the stress tensor is accounted for by adding a term σ'_{ij} to the stress tensor,

$$\pi_{ij} = \sigma_{ij} = p\delta_{ij} + \rho u_i u_j - \sigma'_{ij}. \tag{9.27}$$

Dissipation can only arise from velocity gradients, since any constant term added to the velocity can be removed via a Galilean transformation. The dissipative coefficient coupling the stress tensor to the velocity gradient is most generally a fourth rank tensor,

$$\sigma'_{ij} = \eta_{ijkl} \nabla_k u_l. \tag{9.28}$$

However, symmetry requires that

$$\sigma'_{ij} = \eta \left(\nabla_i u_j + \nabla_j u_i - \frac{2}{3}\delta_{ij} \nabla \cdot \mathbf{u} \right) + \zeta \delta_{ij} \nabla \cdot \mathbf{u}. \tag{9.29}$$

This gives the Navier–Stokes equations. Assuming incompressibility, we have

$$\frac{\partial \rho}{\partial t} = 0 = -\nabla \cdot (\rho \mathbf{u}) = -\nabla \cdot \mathbf{u}. \tag{9.30}$$

Thus,

$$\rho \frac{\partial \mathbf{u}}{\partial t} + \rho (\mathbf{u} \cdot \nabla)\mathbf{u} = -\nabla p + \eta \nabla^2 \mathbf{u}, \tag{9.31}$$

along with the constraint $\nabla \cdot \mathbf{u} = 0$. The velocity field thus has purely transverse components.

How is this to be generalized for a nematic fluid? First, for a nematic fluid, we must have an equation of motion for the director $\mathbf{n}$ (or, equivalently, for the $\mathbf{Q}_{\alpha\beta}$ tensor) in addition to the equations for the conservation of matter, momentum and energy [11,43]. Second, distortions of configurations of the nematic order parameter field also contribute to the stress tensor of the system [11,43].

The director is aligned with the local molecular field in equilibrium; local distortions away from the molecular field direction must relax to minimize the free energy. The local molecular field is defined in the equal Frank constant approximation as

$$h_i = K\nabla^2 n_i. \tag{9.32}$$

Also, the director does not change under rigid translations at constant velocity. Thus, the leading coupling of $\mathbf{n}$ to $\mathbf{u}$ must involve gradients of $\mathbf{n}$. This can then be written as

$$\frac{\partial n_i}{\partial t} - \lambda_{ijk} \nabla_j u_k + X'_i = 0, \tag{9.33}$$

where X' is the dissipative part of the current. This dissipative part can be written as

$$X'_i = \delta^T_{ij} \frac{1}{\gamma} h_j, \tag{9.34}$$

where γ is a dissipative coefficient and the projector isolates components of the fluctuation in the plane perpendicular to the molecular field direction.

The constraint $\mathbf{n} \cdot \partial \mathbf{n}/\partial t = 0$ implies that there are only two independent components of the tensor λ_{ijk}. These can be taken to be symmetric and antisymmetric, defining

$$\lambda_{ijk} = \frac{1}{2}\lambda \left(\delta_{ij}^{\mathrm{T}} n_k + \delta_{ik}^{\mathrm{T}} n_j\right) + \frac{1}{2}\lambda_2 \left(\delta_{ij}^{\mathrm{T}} n_k - \delta_{ik}^{\mathrm{T}} n_j\right), \qquad (9.35)$$

where

$$\delta_{ij}^{\mathrm{T}} = \delta_{ij} - n_i n_j. \qquad (9.36)$$

Under a rigid rotation,

$$\frac{\partial n_i}{\partial t} = \boldsymbol{\omega} \times \mathbf{n} = \frac{1}{2}(\nabla \times \mathbf{u}) \times \mathbf{n}, \qquad (9.37)$$

mandating that the coefficient λ_2 of the antisymmetric part must be -1.

Thus, the final equation of motion for the director, the Oseen equation, takes the form [11,43]

$$\partial_t n_i + \mathbf{v} \cdot \nabla n_i + \omega_{ij} n_j = \delta_{ij}^{\mathrm{T}} \left(\lambda u_{ij} n_j + \frac{h_i}{\gamma}\right). \qquad (9.38)$$

where $u_{ij} = (1/2)(\partial_i u_j + \partial_j u_i)$.

The stress tensor of the nematic consists of three parts. The first is the thermodynamic pressure p, while the second is the viscous stress, given by the tensor

$$\sigma'_{ij} = \alpha_1 n_i n_j n_k n_l A_{kl} + \alpha_2 n_j N_i + \alpha_3 n_i N_j + \alpha_4 A_{ij}$$
$$+ \alpha_5 n_j n_k A_{ik} + \alpha_6 n_i n_k A_{jk}, \qquad (9.39)$$

constructed from symmetry allowed terms, where $\mathbf{A}$ represents the symmetric part of the velocity gradient as before and the vector $N_i = \dot{n}_i + \frac{1}{2}(\hat{\mathbf{n}} \times \mathrm{curl}\mathbf{u})_i$ is the change of director with respect to the background fluid. One relation, due to Parodi, connects the coefficients $\alpha_1 \ldots \alpha_6$; there are thus five independent coefficients of viscosity in the nematic [11].

The third contribution to the stress tensor is the static (elastic) contribution arising from deformations in the director field [11], i.e.,

$$\sigma_{ij}^e = -\frac{\partial F}{\partial \nabla_j n_k} \nabla_i n_k. \qquad (9.40)$$

A more detailed discussion of the equation of motion of the nematic order parameter and the stress tensor is available in [11].

9.4.2 Active Orientational Order and Its Instabilities

The central idea behind the modelling of the active nematic is that out of thermal equilibrium, new terms enter the equation of motion for the nematic director as well as the stress tensor.[1] These terms are more "relevant" than the terms mandated by thermodynamic approaches, in the sense that their effects are stronger at long wavelengths and at long times, i.e., in the thermodynamic limit. To demonstrate this, assume nematic or polar ordered particles, given by a unit director field $\mathbf{n}$. The "slow" or hydrodynamic variables are (i) the concentration fluctuations $\delta c(\mathbf{r}, t)$, (ii) total (solute and solvent) momentum density $\mathbf{g}(\mathbf{r}, t) = \rho \mathbf{u}(\mathbf{r}, t)$ and broken symmetry variables whose fluctuations $\delta n_\perp = \mathbf{n} - \hat{z}$. For polar ordered suspensions, there is a non-zero drift velocity $v_0 \mathbf{n}$. The momentum density evolves via

$$\frac{\partial g_i}{\partial t} = -\nabla_j \sigma_{ij}, \tag{9.41}$$

with σ_{ij} the stress tensor, incorporating pressure and convective terms.

What is special about self-propelled particle systems is that the active contribution to the stress tensor is proportional to the nematic order parameter, i.e.,

$$\sigma_{ij}^{(p)} \propto \left(n_i n_j - \frac{1}{3} \delta_{ij} \right). \tag{9.42}$$

Pascals law is thus violated in the active nematic, since it has a non-zero deviatoric stress. Also, crucially, terms obtained from the standard near-equilibrium analysis are of higher order, since they involve gradients of the nematic order parameter.

This result can be derived from a microscopic calculation using the simple model for a single swimmer discussed above, by simply Taylor expanding the dipole term, which appears in the force density, about the hydrodynamic centre. This yields

$$\sigma_{ij}^a = \frac{a_L + a_S}{2} f c(\mathbf{r}, t) \left(n_i n_j - \frac{1}{3} \delta_{ij} \right), \tag{9.43}$$

where a, a' are the same as the a_L and a_S defined earlier in the context of the single swimmer. We assume an initially aligned state in which the director points, on average, along the $\hat{z}$ direction. Fluctuations away from this are given by $\delta \mathbf{n}_\perp = \hat{n} - \hat{z}$. Fluctuations about this averaged value can be parameterized by

$$\delta \mathbf{n} = \begin{pmatrix} \delta n_x \\ \delta n_y \\ 1 \end{pmatrix}. \tag{9.44}$$

---

[1] In this section, we follow the treatments of [15, 44] closely.

Also, we can expand the velocity field **u** in terms of fluctuations about its mean value u_0,

$$\delta\mathbf{u} = \begin{pmatrix} \delta u_x \\ \delta u_y \\ u_0 + \delta u_z \end{pmatrix}. \tag{9.45}$$

Finally, we must allow for concentration fluctuations, via

$$c(\mathbf{r}, t) = c_0 + \delta c(\mathbf{r}, t). \tag{9.46}$$

Using the equation of motion for the velocity field, defined through $\mathbf{u} = \mathbf{g}/\rho$, we get

$$\partial_t \mathbf{u} = -\nabla_j \left[\frac{(a_L + a_S)}{2} \left(\frac{f}{\rho}\right)(c_0 + \delta c(\mathbf{r}, t)) \left(n_i n_j - \frac{1}{3}\delta_{ij}\right) \right]. \tag{9.47}$$

This then gives, once we insert the expressions for small fluctuations and retain terms to the lowest order

$$\partial_t \delta u_x = \frac{(a_L + a_S)}{2}\left(\frac{f}{\rho}\right) c_0 \partial_z \delta n_x$$

$$\partial_t \delta u_y = \frac{(a_L + a_S)}{2}\left(\frac{f}{\rho}\right) c_0 \partial_z \delta n_y$$

$$\partial_t \delta u_z = \frac{(a_L + a_S)}{2}\left(\frac{f}{\rho}\right) c_0 (\partial_x \delta n_x + \partial_y \delta n_y) + \frac{(a_L + a_S)}{2}\left(\frac{f}{\rho}\right) \partial_z \delta c(\mathbf{r}, t), \tag{9.48}$$

which we can write as

$$\partial_t \delta\mathbf{u}_\perp \simeq w_0 \partial_z \delta\mathbf{n}_\perp$$

$$\partial_t \delta u_z \simeq w_0 \nabla_\perp \cdot \delta\mathbf{n}_\perp + \alpha \partial_z \delta c(\mathbf{r}, t), \tag{9.49}$$

where $\alpha \simeq fa/\rho$ and $w_0 \simeq c_0 a$ and $a \simeq \frac{(a_L + a_S)}{2}$. We must now impose incompressibility, which requires that

$$\nabla_\perp \delta\mathbf{u}_\perp + \partial_z u_z = 0. \tag{9.50}$$

In Fourier space, this is

$$i\mathbf{q}_\perp \cdot \delta\mathbf{u}_\perp(\mathbf{q}, t) + i q_z u_z(q_z, t) = 0. \tag{9.51}$$

Incompressibility implies that the velocity field must be purely transverse, a condition that is easily imposed by using the transverse projection operator

$$P_{\alpha\beta}^T = \left(\delta_{\alpha\beta} - \frac{q_\alpha q_\beta}{q^2}\right). \tag{9.52}$$

Operating this projection operator on a vector field isolates those parts of the field, which have no longitudinal component. Doing this yields

$$\partial_t \delta u_x = -i w_0 \left(1 - \frac{q_x^2}{q^2}\right) q_z \delta n_x - \frac{q_x q_y}{q^2} q_z \delta n_y - i\alpha \frac{q_x q_z}{q^2} \left(q_x \delta n_x + q_y \delta n_y + q_z \delta c\right)$$

$$\partial_t \delta u_y = -i w_0 \left(-\frac{q_x q_y}{q^2}\right) q_z \delta n_x + \left(1 - \frac{q_y^2}{q^2}\right) q_z \delta n_y$$

$$-i\alpha \frac{q_x q_z}{q^2}(q_x \delta n_x + q_y \delta n_y + q_z \delta c), \tag{9.53}$$

which can be written in the compact form

$$\frac{\partial \delta \mathbf{u}_\perp}{\partial t} = -i w_0 q_z \left(1 - 2\mathbf{q}_\perp \mathbf{q}_\perp / q^2\right) \delta \mathbf{n}_\perp - i\frac{q_z^2}{q^2}\alpha(\mathbf{q}_\perp \delta c). \tag{9.54}$$

Number conservation implies

$$\frac{\partial \delta c}{\partial t} = -\nabla \cdot \mathbf{j}, \tag{9.55}$$

where $\mathbf{j} = c v_0 \mathbf{n} = c v_0 (\delta \mathbf{n}_\perp + \hat{z})$, so

$$\left(\frac{\partial}{\partial t} + v_0 \frac{\partial}{\partial z}\right) \delta c + c_0 v_0 \nabla_\perp \cdot \delta \mathbf{n} = 0. \tag{9.56}$$

The equation of motion for polar-ordered particles whose orientation is described by $\mathbf{n}$ contains a term representing advection by a mean drift, a term describing the consequences of a non-equilibrium osmotic pressure and other terms familiar from our brief description of nematodynamics close to equilibrium in the previous section [44],

$$\frac{\partial \delta \mathbf{n}_\perp}{\partial t} = -\lambda_1 v_0 \frac{\partial \delta \mathbf{n}_\perp}{\partial z} - \sigma_1 \nabla_\perp \delta c + \frac{1}{2}\partial\left(\frac{\delta \mathbf{u}_\perp}{\partial z} - \nabla_\perp u_z\right) + \frac{1}{2}\gamma_2 \left(\frac{\partial \delta \mathbf{u}_\perp}{\partial z} + \nabla_\perp u_z\right). \tag{9.57}$$

Taking the curl of this equation as well as the equation for $\mathbf{u}_\perp$ gives coupled equations for the dynamics of twist $\nabla \times \mathbf{n}_\perp$ and/or vorticity $\nabla \times \mathbf{u}_\perp$. Related results follow from taking the divergence of these two equations, resulting in coupled equations for $\nabla \cdot \mathbf{n}_\perp$, $\nabla \cdot \mathbf{u}_\perp$ and $\nabla \delta c$, which are complicated but have also been analyzed [44]. These coupled equations have been shown to possess wave-like solutions.

The results of these calculations are summarized in the following:

- A linearized treatment, ignoring viscosity, for the polar or apolar cases at lowest order in wave number, yields propagating modes with a characteristic instability in the case of purely apolar active particles.

- Retaining viscosity, in the steady Stokesian limit where accelerations are ignored, polar and nematic orders at small wave numbers are generically destabilized by a coupling of splay (for contractile particles) or bend (for tensile particles) modes to the hydrodynamic velocity field. In this limit, for nematic order, this instability has been referred to as a generic instability.
- Any mechanism for introducing screening into the hydrodynamic interaction will suppress the instability, such as introducing boundaries into the system [6, 39] or applying a shear [30].
- Number fluctuations in ordered collections of self-propelled particles are anomalously large [46, 47]. The variance $\langle (\delta N)^2 \rangle$, scaled by the mean N, diverges as $N^{2/3}$ in three dimensions in the linearized treatment of [44]. Physically, this is a consequence of the fact that (a) orientational fluctuations (director distortions) produce mass flow and (b) such fluctuations are large because director fluctuations arise from a broken symmetry mode.

In a more general context, activity provides new currents both for matter and momentum, beyond those which would be predicted from a theory based on small perturbations away from thermodynamic equilibrium. It is these new currents which are the source of the novel results indicated above.

9.4.3 Rheological Predictions

The following predictions for the rheological properties of active nematics are obtained from [15] and the discussion here follows that reference closely; see also [1, 22, 24]. The active contribution to force densities within the fluid follows from

$$F^a = \nabla \cdot \sigma^a. \tag{9.58}$$

The activity contributes to traceless symmetric (deviatoric) stress through

$$\sigma^a - \frac{1}{3} Tr \sigma^a \simeq W \mathbf{Q} + W_2 \mathbf{Q}^2, \tag{9.59}$$

where W_1 and W_2 reflect the strength of the elementary dipoles. The full stress tensor has contributions from the fluid (σ^v) as well as the order parameter field (σ^{OP}), giving

$$\sigma = \sigma^a + \sigma^v + \sigma^{OP}. \tag{9.60}$$

The time-dependence of σ is assumed to be slaved to the time-dependence of the order parameter field $\mathbf{Q}$. The equation of motion for $\mathbf{Q}$, upon linearization, should take the form

$$\frac{\partial \mathbf{Q}}{\partial t} = -\frac{1}{\tau} \mathbf{Q} + D \nabla^2 \mathbf{Q} + \lambda_0 \mathbf{A} + \dots, \tag{9.61}$$

where τ is an activity correlation time, D is a diffusivity, λ_0 is a kinetic coefficient and higher order terms have been dropped in favor of the lowest-order ones. Using these, we can calculate the stress response to a shear in the xy plane. In Fourier space, this is

$$\sigma_{xy}(\omega) = -\left[\eta_0 + \frac{(a + W)\lambda_0}{-i\omega + \tau^{-1}}\right] A_{xy}$$

$$= -\frac{G'(\omega) - iG''(\omega)}{\omega} i A_{xy}, \tag{9.62}$$

defining the storage and loss moduli $G'(\omega)$ and $G''(\omega)$. The important results which follow from this analysis are the active enhancement or reduction $\eta_{act} \propto W\tau$ (depending on the sign of the parameter W) of the effective viscosity at zero-shear rate. The theory also predicts strong viscoelasticity as τ increases. For passive systems $W = 0$. For active systems, W is non-zero and the storage modulus then behaves as

$$G'(\omega\tau \gg 1) \simeq W. \tag{9.63}$$

All these effects are expected to be enhanced if the transition is to a polar ordered phase, rather than to a nematic.

9.5 Active Gels: Summary

A parallel line of activity, centered on models for specific biological phenomena, such as the symmetry-breaking motility exhibited by beads coated with polymerizing actin and the dynamical topological defect structures obtained in mixtures of motors and microtubules, outlines a description of active matter in terms of *active gels* [13, 17, 19, 20] .

The philosophy of these approaches is the following: Rather than begin from a microscopic model for a swimmer or individual moving particle and then generalize from the microscopics to realize symmetry-allowed equations of motion for the fluid velocity field and for the local concentration of swimmers, one can start with a coarse-grained continuum model for a physical viscoelastic gel which is driven by internally generated, non-equilibrium sources of energy. The equations of motion for the stresses in this gel as well as for local order-parameter-like quantities describing, for example, polar order at a coarse-grained scale are constructed using the basic symmetries of the problem. The non-equilibrium character of this problem follows from the fact that such equations of motion do not derive from an underlying free energy.[2]

The passive gel has a viscoelasticity whose simplest representation is via a Maxwell model, exhibiting solid-like behavior at short times and fluid-like behavior

---

[2] A summary of these results can be found in [17], from where most of this material is drawn.

at long times. In this model, the deviatoric stress $\sigma_{\alpha\beta}$ is related to the strain rate tensor $u_{\alpha\beta} = \frac{1}{2}\left(\partial_\alpha u_\beta + \partial_\beta u_\alpha\right)$, where $\mathbf{u}$ is the velocity field in the gel, via

$$\frac{\partial \sigma_{\alpha\beta}}{\partial t} + \frac{\sigma_{\alpha\beta}}{\tau} = 2E u_{\alpha\beta}, \tag{9.64}$$

where E is a shear modulus obtained at short times. The simple time derivative must be augmented by convective terms, as well as terms representing the effects of local rotation of the fluid, to enforce Galilean invariance and the appropriate *frame independence*.

Polar order in such gels is assumed to be weighted by a free-energy-like expression obtained from the theory of polar nematic liquid crystals. This takes the form:

$$\mathscr{F} = \int d\mathbf{r}\left[\frac{K_1}{2}(\nabla \cdot \mathbf{p})^2 + \frac{K_2}{2}(\mathbf{p} \cdot (\nabla \times \mathbf{p}))^2 + \frac{K_3}{2}(\mathbf{p} \times (\nabla \times \mathbf{p}))^2 + k\nabla \cdot \mathbf{p} - \frac{h_\parallel^0}{2}\mathbf{p}^2\right].$$

$$\tag{9.65}$$

There are three Frank constants for splay, twist and bend, as in the nematic case. The (non-zero) constant k is permitted by the vectorial symmetry of the polar case. The amplitude of local order is parametrized by the constant $h_\parallel^0$.

The hydrodynamic theory of active gels begins by identifying, along classical lines, fluxes and forces. The hydrodynamic description contains phenomenological parameters, called Onsager coefficients. These fluxes are the mechanical stress $\sigma_{\alpha\beta}$ associated with the mechanical behavior of the cell, the rate of change of polar order (the polarization) $\dot{\mathbf{P}}$ and the rate of consumption of ATP per unit volume r. The generalized force conjugate to the ATP consumption rate is the chemical potential difference $\Delta\mu$ between ATP and the products of ATP hydrolysis, while the force conjugate to changes in the polarization is the local field $\mathbf{h}$, obtained from the functional derivative of the free energy, i.e., $\mathbf{h} = -\delta\mathscr{F}/\delta\mathbf{p}$. The force conjugate to the stress tensor is, as usual, the velocity gradient tensor $\partial_\alpha u_\beta$. This can, as is conventionally done, be expanded into its traceless symmetric, pure trace and antisymmetric parts. A similar expansion can be made for the stress tensor.

The next step is to construct equations of motion for the deviatoric stress, using the convected Maxwell model with a single viscoelastic relaxation time. The equation must couple the mechanical stress and the polarization field as well as include a term coupling activity to the stress. It takes the form

$$2\eta u_{\alpha\beta} = \left(1 + \tau\frac{D}{Dt}\right)\left\{\sigma_{\alpha\beta} + \zeta\Delta\mu q_{\alpha\beta} + \tau A_{\alpha\beta}\right.$$
$$\left. - \frac{\nu_1}{2}\left(p_\alpha h_\beta + p_\beta h_\alpha - \frac{2}{3}h_\gamma p_\gamma \delta_{\alpha\beta}\right)\right\}, \tag{9.66}$$

where the co-rotational derivative is

$$\frac{D}{Dt}\sigma_{\alpha\beta} = \left(\frac{\partial}{\partial t} + u_\gamma\frac{\partial}{\partial r_\gamma}\right)\sigma_{\alpha\beta} + \left[\omega_{\alpha\gamma}\sigma_{\gamma\beta} + \omega_{\beta\gamma}\sigma_{\gamma\alpha}\right]; \tag{9.67}$$

the tensor $A_{\alpha\beta}$ describes geometrical nonlinearities arising out of generalizations of the Maxwell model to viscoelastic fluids, and $q_{\alpha\beta} = \frac{1}{2}\left(p_\alpha p_\beta - \frac{1}{3}p^2 \delta_{\alpha\beta}\right)$. The antisymmetric part of the stress tensor leads to torques on the fluid and is obtainable from

$$\sigma^a_{\alpha\beta} = \frac{1}{2}\left(p_\alpha h_\beta - p_\beta h_\alpha\right). \tag{9.68}$$

The viscoelastic relaxation time is τ, the co-efficient ν_1 describes the coupling between mechanical stresses and the polarization field, while the parameter ζ is the co-efficient of active stress generation, acting to couple activity to the stress. The second flux, defined from the rate of change of polarization is given by

$$\dot{\mathbf{P}} = \frac{D\mathbf{P}}{Dt}. \tag{9.69}$$

The Onsager relation for the polarization is

$$\frac{D}{Dt}p_\alpha = \frac{1}{\gamma_1}h_\alpha + \lambda_1 p_\alpha \Delta\mu - \nu_1 u_{\alpha\beta} p_\beta - \bar{\nu}_1 u_{\beta\beta} p_\alpha. \tag{9.70}$$

These include several phenomenological parameters, such as the rotational viscosity γ_1, which characterizes dissipation from the rotation of the polarization as well as the constants ν_1 and $\bar{\nu}_1$.

Then, we must have an equation for the rate of consumption of ATP. This takes the form

$$r = \Lambda\Delta\mu + \zeta p_\alpha p_\beta u_{\alpha\beta} + \bar{\zeta} u_{\alpha\alpha} + \lambda_1 p_\alpha h_\alpha. \tag{9.71}$$

These are simple but generic equations representing the basic symmetries of the problem. They can be shown to have interesting and surprising consequences: an active polar gel can exhibit spontaneous motion as a consequence of a gradient in the polar order parameter, as well as defects in the polar ordering, which are dynamic in character [17]. These ideas have been applied to the study of the motion of the cell lamellipodium and to the organization of microtubules by molecular motors.

The generality of these equations follows from the fact that they are motivated principally by symmetry considerations. Thus, even though they describe intrinsically non-equilibrium and highly nonlinear phenomena, for which the rules for constructing effective, coarse-grained equations of motion for the basic fields are not as well developed as the theory for the relaxation of small perturbations about thermal equilibrium, the "unreasonable effectiveness of hydrodynamics" may well hold in their favor.

9.6 Conclusions

This chapter has provided a brief review of the field of what is currently called active matter. The emphasis has been on attempting to clarify the basic ideas which have motivated the development of this field, rather than the details of the often intricate and complex calculations implementing these ideas. Much recent and important work, including numerical calculations – illustrative references are [28, 36, 40] – has been omitted entirely for the sake of compactness.

As indicated in the introduction, the importance of this field would appear to be that it might suggest ways of thinking about the response and dynamics of living systems, while providing a largely self-consistent framework for calculations. Several non-trivial insights have already been obtained from these calculations, particularly in the identification of the generic instability of polar or nematically ordered states in the presence of the long-ranged hydrodynamic interaction, the connection between microscopic models and their hydrodynamic limits, as well as a comprehensive theory of active gels, generalizing ideas from nematic physics. The precise relationship between macroscopic, symmetry-based hydrodynamic equations representing active nematics and an underlying microscopic theory has been substantially clarified, as in the work of [6] and references cited therein. To what extent further developments in this field may aid the increasingly *active* dialogue between physics and the engineering sciences on the one hand and the biological sciences on the other remains to be seen.

Acknowledgements I thank Sriram Ramaswamy and Madan Rao for many enlightening and valuable discussions concerning the physics of active matter. Conversations at various points of time with Cristina Marchetti, Tanniemola Liverpool, Jacques Prost, Karsten Kruse, Frank Julicher, David Lacoste, Ronojoy Adhikari, P. B. Sunil Kumar, Aparna Baskaran and Jean-Francois Joanny have also helped to shape the material in this chapter. This work was supported by DST (India) and by the Indo-French Centre for the Promotion of Advanced Research (CEFIPRA) [Grant No. 3502]. The hospitality of ESPCI and the Institut Henri Poincare is gratefully acknowledged.

References

1. Ahmadi A, Marchetti MC, Liverpool TB (2006) Hydrodynamics of isotropic and liquid crystalline active polymer solutions. Phys Rev E 74:061913
2. Aranson IS, Tsimring LS (2005) Pattern formation of microtubules and motors: inelastic interaction of polar rods. Phys Rev E 71:050901(R)
3. Ballerini M, Cabibbo N, Candelier R et al (2008) Interaction ruling animal collective behavior depends on topological rather than metric distance: evidence from a field study. Proc Nat Acad Sci 105:1232–1237
4. Baskaran A, Marchetti MC (2008) Enhanced diffusion and ordering of self-propelled rods. Phys Rev Lett 101:268101
5. Baskaran A, Marchetti MC (2008) Hydrodynamics of self-propelled hard rods. Phys Rev E 77:011920
6. Baskaran A, Marchetti MC (2009) Statistical mechanics and hydrodynamics of bacterial suspensions. Proc Nat Acad Sci 106:15567–15572
7. Bazazi S, Buhl J, Hale et al (2008) Collective motion and cannibalism in locust migratory bands. Curr Biol 18(10):735–739
8. Bershadsky A, Kozlov M, Geiger B (2006) Adhesion-mediated mechanosensitivity: a time to experiment, and a time to theorize. Curr Opin Cell Biol 18:472–81
9. Bray D (1992) Cell Movements: from molecules to motility. Garland Publishing, New York
10. Buhl J, Sumpter DJT, Couzin ID et al (2006) From disorder to order in marching locusts. Science 312(5778):1402–1406

11. de Gennes PG, Prost J (1993) Physics of liquid crystals, 2nd edn. Clarendon, Oxford
12. Dombrowski C, Cisneros L, Chatkaew L et al (2004) Self-concentration and large-scale coherence in bacterial dynamics. Phys Rev Lett 93:098103
13. Giomi L, Marchetti MC, Liverpool TB (2008) Complex spontaneous flows and concentration banding in active polar films. Phys Rev Lett 101:198101
14. Gruler H, Dewald U, Eberhardt M (1999) Nematic liquid crystals formed by living amoeboid cells. Eur Phys J B 11:187–192
15. Hatwalne Y, Ramaswamy S, Rao M et al (2004) Rheology of active-particle suspensions. Phys Rev Lett 92:118101
16. Joanny JF, Prost J (2008) Active gels as a description of the actin-myosin cytoskeleton. HFSP J 3(2):94–104
17. Jülicher F, Kruse K, Prost et al (2007) Active behavior of the cytoskeleton. Phys Rep 449:3–28
18. Karsenti E, Nedelec F, Surrey T (2006) Modeling microtubule patterns. Nat Cell Biol 8:1204–1211
19. Kruse K, Joanny JF, Jülicher F et al (2004) Asters, vortices, and rotating spirals in active gels of polar filaments. Phys Rev Lett 92:078101
20. Kruse K, Joanny JF, Jülicher F et al (2005) Generic theory of active polar gels: a paradigm for cytoskeletal dynamics. Eur Phys J E 16:5–16
21. Lacoste D, Menon GI, Bazant MZ et al (2009) Electrostatic and electrokinetic contributions to the elastic moduli of a driven membrane. Eur Phys J E 28:243–264
22. Lau AWC, Lubensky T (2009) Fluctuating hydrodynamics and microrheology of a dilute suspension of swimming bacteria. Phys Rev E 80:011917
23. Lee HY, Kardar M (2001) Macroscopic equations for pattern formation in mixtures of microtubules and molecular motors. Phys Rev E 64:056113
24. Liverpool TB, Marchetti MC (2006) Rheology of active filament solutions. Phys Rev Lett 97:268101
25. Makris NC, Ratilal P, Symonds DT et al (2000) Fish population and behavior revealed by instantaneous continental shelf-scale imaging. Science 311:660–663
26. Manneville JB, Bassereau P, Ramaswamy S, Prost J (2001) Active membrane fluctuations studied by micropipet aspiration. Phys Rev E 64:021908
27. Manneville JB, Bassereau P, Levy D, Prost J (1999) Activity of Transmembrane Proteins Induces Magnification of Shape Fluctuations of Lipid Membranes. Phys Rev Lett 82:4356
28. Marenduzzo D, Orlandini E, Cates ME et al (2008) Lattice Boltzmann simulations of spontaneous flow in active liquid crystals: the role of boundary conditions. J Non-Newt Fluid Mech 149:56–62
29. Marcy Y, Prost J, Carlier MF et al (2004) Forces generated during actin-based propulsion: a direct measurement by micromanipulation. Proc Nat Acad Sci 101:5992–5997
30. Muhuri S, Rao M, Ramaswamy S (2007) Shear-flow-induced isotropic to nematic transition in a suspension of active filaments. Europhys Lett 78:48002
31. Narayanan V, Ramaswamy S, Menon N (2007) Long-lived giant number fluctuations in a swarming granular nematic. Science 317:105–108
32. Nedelec FJ, Surrey T, Maggs AC et al (1997) Self-organization of microtubules and motors. Nature 389:305–308
33. Prost J, Bruinsma R (1996) Shape fluctuations of active membranes. Europhys Lett 33:321
34. Purcell EM (1977) Life at low Reynolds number. Am J Phys 45:3–11
35. Rafelski SM, Theriot JA (2004) Crawling toward a unified model of cell motility: spatial and temporal regulation of actin dynamics. Ann Rev Biochem 73:209–239
36. Ramachandran S, Sunil Kumar PB, Pagonabarraga I (2006) A Lattice-Boltzmann model for suspensions of self-propelling colloidal particles Eur Phys J E 20:151–158
37. Ramaswamy S, Simha RA, Toner J (2003) Active nematics on a substrate: giant number fluctuations and long-time tails. Europhys Lett 62:196–202
38. Ramaswamy S, Toner J, Prost J (2000) Nonequilibrium fluctuations, traveling waves, and instabilities in active membranes. Phys Rev Lett 84:3494
39. Ramaswamy S, Rao M (2007) Active filament hydrodynamics: instabilities, boundary conditions and rheology. New J Phys 9:423

40. Saintillan D, Shelley MJ (2008) Instabilities and pattern formation in active particle suspensions: kinetic theory and continuum simulations. Phys Rev Lett 100:178103
41. Sankararaman S, Menon GI, Kumar PBS (2004) Self-organized pattern formation in motor-microtubule mixtures (2004) Phys Rev E 70:031905
42. Sankararaman S, Menon GI, Kumar PBS (2002) Two-component fluid membranes near repulsive walls: linearized hydrodynamics of equilibrium and nonequilibrium states. Phys Rev E 66:031914
43. Stark H, Lubensky TC (2003) Poisson-bracket approach to the dynamics of nematic liquid crystals. Phys Rev E 76:061709
44. Simha RA, Ramaswamy S (2002) Hydrodynamic fluctuations and instabilities in ordered suspensions of self-propelled particles. Phys Rev Lett 89:058101
45. Simha RA, Ramaswamy S (2002) Statistical hydrodynamics of ordered suspensions of self-propelled particles: waves, giant number fluctuations and instabilities. Phys A 306:262–269
46. Toner J, Tu Y (1998) Long-range order in a two-dimensional dynamical XY model: how birds fly together. Phys Rev Lett 75:4326–4329
47. Toner J, Tu Y (1998) Flocks, herds, and schools: a quantitative theory of flocking. Phys Rev E 58:4828–4858
48. Toner J, Tu Y, Ramaswamy S (2005) Hydrodynamics and phases of flocks. Ann Phys 318:170–244
49. Turner L, Ryu WS, Berg HC (2000) Real-Time Imaging of Fluorescent Flagellar Filaments. J Bacteriol 182:2793–2801
50. Verkhovsky AB, Svitkina TM, Borisy GG (1999) Self-polarization and directional motility of cytoplasm. Curr Biol 9(1):11–20
51. Vicsek T, Czirk A, Ben-Jacob E et al (1995) Novel type of phase transition in a system of self-driven particles. Phys Rev Lett 75:1226–1229

Chapter 10
Mathematical Modelling of Granular Materials

Mehrdad Massoudi

Abstract In this chapter, we provide a brief overview of the important issues in modelling of granular materials. A continuum mechanics approach is taken where it is assumed that the material behaves similar to a compressible non-linear fluid where the effects of density gradients are incorporated in the stress tensor. We discuss and solve the heat transfer in granular materials flowing down an inclined plane with radiation effects at the free surface. For a fully developed flow, the equations simplify to a system of three non-linear ordinary differential equations. The equations are made dimensionless and a parametric study is performed where the effects of various dimensionless numbers representing the effects of heat conduction, viscous dissipation, radiation, etc. are presented.

10.1 Introduction

A great many industrial processes involve interaction between (solid) particles and fluids. Some of these processes are: combustion, fluidization, coal gasification, drying of particles, catalytic and thermal cracking, heat exchangers. A recent study indicates that the commercial and large-scale solids-processing plants have an average operating reliability of 63%, compared to 84% for large-scale plants using only liquids and gases [80]. A major challenge facing the designers of coal gasification plants is to assure reliable and efficient movement of solids into and out of high-pressure, high-temperature fluidized-bed processing units. Earlier studies of the flow of granular material were mainly concerned with the engineering and structural design of bins and silos. The inaccuracy of these theories, especially for dynamic conditions of loading or emptying, occasionally resulted in failure of the bin or silo.

M. Massoudi (✉)
U.S. Department of Energy, National Energy Technology Laboratory (NETL),
Pittsburgh, PA 15236, USA
e-mail: massoudi@netl.doe.gov

J.M. Krishnan et al. (eds.), *Rheology of Complex Fluids*,
DOI 10.1007/978-1-4419-6494-6_10, © Springer Science+Business Media, LLC 2010

In a typical fossil-based fuel power plant, shown in Fig. 10.1, the solid particles, e.g. coal, after having been cleaned and processed are usually stored in some configurations, such as Regimes A and/or B. In Regime B, the granular particles are contained in a bin or hopper, and depending on the geometry and the application, the particles undergo a slow frictional flow whereby either the Mass Flow or the Funnel Flow occurs [93]. In Regime C, where many of the handling of bulk solids, such as pharmaceutical [48], agricultural and drying processes occur, the particles are colliding with each other and bouncing on top of each other; this flow regime is known as the Rapid Flow Regime. (It should be noted that the transition regime (also called the intermediate region), which occurs in many applications, probably involves the most difficult problems to study in granular materials. For an excellent discussion of this issue, see [113]). In Regime D, the granules are transported to the combustion chamber by mixing them with either a gas (air), known as the pneumatic transport [55], or a liquid (water or oil), known as the hydraulic transport of particles [106]. This regime is where coal slurries are modelled either as a suspension flow using non-Newtonian fluid models, or as mixtures or multi-phase models [89]. In Regime E, the fuel (mixed solids and air or other gases) is burnt and this regime is known as the Chemically Reactive Flow Regime [25, 28, 44, 92]. In Regimes F and G, the interest shifts either to the dispersed particles (or pollutants) released into atmosphere, or solid wastes usually collected or processed at this stage, i.e., Regime G. It can be seen that at some point, i.e., in certain flow regimes such as A, B or C, one can model the materials using single phase continuum approach, or use statistical or numerical methods to study the flow and behaviour of these particles. In flow Regime D, it is perhaps more appropriate to model the flow using the dense multi-phase approach or methods of Suspension Rheology, and in flow Regimes E and F, the dilute (or dispersed) phase approach is often used [65].

In this chapter, we are mainly concerned with mathematical modelling of granular materials in the dense flow regimes (Regime B and C shown in Fig. 10.1). A granular medium is not a solid continuum since it deforms to the shape of the vessel containing it; it cannot be considered a liquid for it can be piled into heaps; and it is not a gas for it will not expand to fill the vessel containing it. A granular material is defined as an assembly of discrete solid components dispersed in space such that the solid constituents are in contact with the near neighbours. The behaviour of granular materials is governed by inter-particle cohesion, friction, collisions, etc. Traditionally, a granular material is assumed to include the combined range of granular powders and granular solids with components ranging in size from about $10 \, \mu m$ up to $3 \, mm$. A powder is composed of particles up to $100 \, \mu m$ (diameter) with further subdivision into ultrafine ($0.1-1.0 \, \mu m$), superfine ($1-10 \, \mu m$) or granular ($10-100 \, \mu m$) particles. A granular solid consists of materials ranging from about 100 to $3,000 \, \mu m$ [13].

Granular materials present a special challenge to engineers and scientists. It is a multidisciplinary field. Experimental studies in soil mechanics dominated this field for a long time. Until a few decades ago where theoretical studies based on modern continuum mechanics started to be used, there were very few mathematical theories of granular materials. One of the outstanding issues in this field is the constitutive

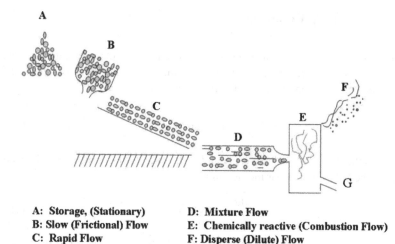

A: Storage, (Stationary)   D: Mixture Flow
B: Slow (Frictional) Flow   E: Chemically reactive (Combustion Flow)
C: Rapid Flow   F: Disperse (Dilute) Flow
                                  G: Solid waste

Fig. 10.1   Various possible flow conditions in a typical coal-based power plant

modelling of the stress tensor. Since this covers a wide rage of disciplines from glaciology [39] to soil mechanics [51], from dilute flow of particles to dense flow applications – it is very difficult to obtain (or derive) a single constitutive relation, which is acceptable and applicable to all these different flow conditions and regimes. In fact, study of granular materials has become part of a study of a larger field of materials with microstructure. One can try to understand the mechanics (or physics) of granular materials either by performing simple experiments to characterize their behaviour or by trying to formulate a theory where the behaviour or the response of granular materials can be studied. Recent review articles by Savage [103], Hutter and Rajagopal [42], and de Gennes [22], and books by Nedderman [74], Mehta [71], Duran [23], Antony et al. [4] and Rao and Nott [93] address many of the interesting issues in the field of granular materials. In the theoretical approach, there are fundamentally two different and distinct, yet related methods that can be used: one is the statistical theories and the other the continuum theories. In the statistical theories, one can take a particle dynamics approach or one can use (modified) forms of the kinetic theory of gases [15]. These two techniques, especially with the advent of faster and more efficient computers, have become very popular in the last two decades. There are many recent review articles which deal with the statistical theories [36, 37, 43] and kinetic theories as applied to granular materials [30, 62].

  In this chapter, we will focus our attention on continuum theories of granular materials. In Sect. 10.2, the governing equations are given, and in Sect. 10.3, a brief description of the conceptual framework used to derive constitutive relations for granular materials is discussed. In Sect. 10.4, an overview of certain important features of modelling of granular materials is discussed; this forms the core of this chapter. And finally, in Sect. 10.5, a simple boundary value problem is solved, followed by some concluding remarks.

10.2   Governing Equations

The balance laws, in the absence of chemical reactions and electromagnetic effects, are the conservation of mass, linear momentum, angular momentum, and energy [117]. The conservation of mass in the Eulerian form is given by

$$\frac{\partial \rho}{\partial t} + \operatorname{div}(\rho \mathbf{u}) = 0, \tag{10.1}$$

where $\mathbf{u}$ is the velocity, ρ is the density and $\partial/\partial t$ is the partial derivative with respect to time. The balance of linear momentum is

$$\rho \frac{d\mathbf{u}}{dt} = \operatorname{div} \mathbf{T} + \rho \mathbf{b}, \tag{10.2}$$

where d/dt is the material time derivative, $\mathbf{b}$ is the body force and $\mathbf{T}$ is the Cauchy stress tensor. The balance of angular momentum (in the absence of couple stresses) yields the result that the Cauchy stress is symmetric. The energy equation is

$$\frac{d\varepsilon}{dt} = \mathbf{T} \cdot \mathbf{L} - \operatorname{div} \mathbf{q} + \rho r, \tag{10.3}$$

where ε denotes the specific internal energy, $\mathbf{q}$ is the heat flux vector, r is the radiant heating and $\mathbf{L}$ is the velocity gradient. Thermodynamical considerations require the application of the second law of thermodynamics or the entropy inequality. The local form of the entropy inequality is given by [52]:

$$\rho \dot{\eta} + \operatorname{div} \boldsymbol{\varphi} - \rho s \geq 0, \tag{10.4}$$

where $\eta(x, t)$ is the specific entropy density, $\boldsymbol{\varphi}(\mathbf{x}, t)$ is the entropy flux, and s is the entropy supply density due to external sources, and the dot denotes the material time derivative. If it is assumed that

$$\boldsymbol{\varphi} = \frac{1}{\theta} \mathbf{q}, \tag{10.5a}$$

$$s = \frac{1}{\theta} r, \tag{10.5b}$$

where θ is the absolute temperature, then (10.4) reduces to the Clausius–Duhem inequality

$$\rho \dot{\eta} + \operatorname{div} \frac{\mathbf{q}}{\theta} - \rho \frac{r}{\theta} \geq 0. \tag{10.6}$$

For a complete study of a thermomechanical problem, the second law of thermodynamics has to be considered [16, 52, 72, 123]. In other words, in addition to other principles in continuum mechanics such as material symmetry, frame indifference, etc., the second law also imposes certain restrictions on the type of motion and/or the constitutive parameters. (For a thorough discussion of important concepts in constitutive equations of mechanics, we refer the reader to the books by Antman [3], Maugin [66], Coussot [18] and Batra [8]). Since there is no general agreement on the functional form of the constitutive relation and since the Helmholtz free energy is not known, a complete thermodynamical treatment of the present model used in our studies is lacking. In general, the application of the second law of thermodynamics, i.e., the Clausius–Duhem inequality, is a subject matter which has caused some controversy. As pointed out recently, in an important paper by Rajagopal and Srinivasa [91], the usual approach is "to posit some phenomenological law for the fluxes and see what restrictions are imposed on the phenomenological coefficients by the non-negativity of the rate of entropy production." This approach has been used successfully in many problems, but as pointed out by Rajagopal and Srinivasa [91], based on observation, "materials are characterized not only by how they store energy (as indicated by the equation of state) but also by how they produce entropy." In this problem, instead, we use the results of Rajagopal et al. [87, 88] to obtain restrictions on the material parameters, while using simple and reasonable expressions for the undetermined coefficients. To "close" the governing equations, constitutive relations are needed for $\mathbf{T}$, $\mathbf{q}$ and r.

10.3   Conceptual Framework

The volume fraction field $v(\mathbf{x}, t)$ plays a major role in many of the proposed continuum theories of granular materials. That is, even though we talk of distinct solid particles with a certain diameter or shape, in this theory, the particles through the introduction of the volume fraction field have been homogenized, as shown in Fig. 10.2 (See [17] for a thorough discussion of homogenizing the microstructure of granular media). In other words, it is assumed that the material properties of the ensemble are continuous functions of position and time. That is, the material may be divided indefinitely without losing any of its defining properties. A distributed volume

$$V_t = \int v \, dV, \tag{10.7}$$

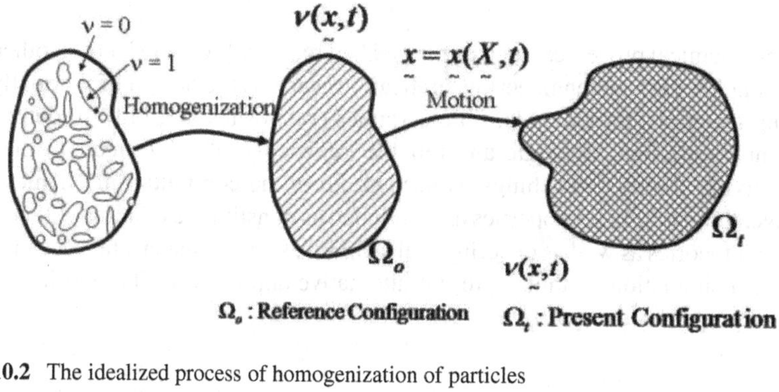

Fig. 10.2 The idealized process of homogenization of particles

and a distributed mass

$$M = \int \rho_s v \mathrm{d}V, \tag{10.8}$$

can be defined, where the function v is an independent kinematical variable called the volume distribution function and has the property

$$0 \le v(\mathbf{x}, t) < v_{\max} < 1. \tag{10.9}$$

The function v is represented as a continuous function of position and time; in reality, v in such a system is either one or zero at any position and time, depending upon whether one is pointing to a granule or to the void space at that position. That is, the real volume distribution content has been averaged, in some sense, over the neighbourhood of any given position. The classical mass density or bulk density, ρ, is related to ρ_s and v through

$$\rho = \rho_s v. \tag{10.10}$$

It is clear that in practice v is never equal to one; its maximum value, generally designated as the maximum packing fraction, depends on the shape, size, method of packing, etc. Furthermore, in most relevant applications in granular media, the particles are assumed to be either spherical or disc-like, and for other irregular shapes, some type of approximation such as a shape factor becomes necessary. However, in certain applications, rod-like or fibrous type granular materials pose further difficulties by introducing a degree of directionality, i.e., anisotropy in the problem. For a discussion of these issues, we refer the reader to a recent paper by Massoudi [58] and the book by Straughan et al. [109]. The introduction of the volume fraction field into the theory to capture, in some sense, the microstructure of the continuum is not the only way that the microstructure (which entails different length scales, shapes, porosity, directionality, etc.) can be accounted for or described. The introduction of the fabric [69, 70, 75, 77, 107] , extension of Cosserat (or micropolar) theories [121] or more advanced continuum theories [2, 26] are among the alternative approaches.

10.4 Constitutive Relations

In many chemical processes, such as packed bed reactors [101, 118], non-isothermal catalyst pellet [76] and engineering applications such as the design of thermally insulating materials, for example, fibre-ceramics composites [53, 114], to determine the temperature field, accurate and reliable knowledge of the (effective) thermal conductivity, among other things, is needed. From the continuum mechanics perspective, these transport properties are modelled as constitutive relations. Of course, statistical theories as well as experimental techniques, and in the last few decades the numerical simulation methods, present alternative approaches. The two important constitutive relations needed for the study of flow and heat transfer in granular materials, where the effects of radiation are ignored, are the stress tensor and the heat flux vector.

Constitutive relations are required to satisfy some general principles. The principle of material frame-indifference (sometimes referred to as objectivity), which requires that the constitutive equations be invariant under changes of frame, is perhaps the most important of all [117]. It is a consequence of a fundamental principle of classical physics that material properties are indifferent, that is, independent of the frame of reference of the observer. This principle requires that constitutive relations depend only on frame-indifferent forms (or combinations thereof) of the variables pertaining to the given problem. Generally speaking, there are two categories in the formulation of constitutive relations where the Principle of Material Frame-Indifference (PMFI) plays a crucial role. The first category is where one has obtained a constitutive relation based on a one-dimensional approximation of an experiment, and then one tries to generalize this correlation to a three-dimensional case. This is more of an issue in co-ordinate transformation, rather than a change of frame. However, if the correlation involves time derivatives, then one has to be more careful. The second category is where the response of the material is timedependent and as a result time derivatives or quantities associated with the movement and rotation of the frame (such as spin) would have to be included in the formulation.

There are many different ways to model the behaviour of granular materials, for example, one can use

- Physical and experimental models
- Numerical simulations
- Statistical mechanics approaches (e.g., extension of kinetic theory of gases)
- Standard continuum mechanics
- Ad-hoc approaches

Within the continuum mechanics approach, it is recognized that granular materials have certain "structures" and as a result "higher" order models or more advanced theories such as micro-mechanics, micropolar, Cosserat theories, non-Newtonian models, hypoplastic or hypoelastic models, viscoelastic, turbulence models, etc. are needed [57]. Another scheme could be the following:

- Explicit constitutive relations
- Implicit constitutive relations
- Ad-hoc relations

An explicit constitutive relation is one where the constitutive variable being modelled is represented "explicitly" in terms of or as a function of other kinematical, thermal, chemical, variables; whereas an implicit constitutive relation is when the constitutive variable, in addition to the above list, somehow depends on its own (time) derivatives, for example. This brings the concept of rate-dependence or rate-independence into the picture. However, since in general it is not clear what is meant by the term "rate-dependent" models, we prefer the explicit and implicit categorization (See [84, 85] for a thorough discussion of this subject).

Historically, many researchers have used the idea of "superposition" of elastic and viscous effects to describe the complex behaviour of materials. Reference [117] gives a review of the subject where ideas of "rate-type", "differential type" or "integral type" materials are discussed. They define a material of differential type as one where the stress $\mathbf{T}$ depends only on a finite number of the time derivatives of $\mathbf{F_t}$. Thus,

$$\mathbf{T}(t) = \sum_{s=0}^{\infty} (\mathbf{F}^t(s)), \tag{10.11}$$

where $\mathbf{F}^t$ is approximated for s near zero, by its Taylor expansion up to some order r such that

$$\mathbf{F}^t(0) = \mathbf{F}(t)$$
$$\dot{\mathbf{F}}^t(0) = \dot{\mathbf{F}}(t)$$
$$\dots \tag{10.12}$$

A *material of rate type* is one where in addition to the general constitutive relation for simple materials,

$$\mathbf{T}(t) = \sum_{s=0}^{\infty} (\mathbf{F}(t - s)), \tag{10.13}$$

we also have a further relationship,

$$\mathbf{T}^{(p)} = \mathbf{g}(\mathbf{T}, \dot{\mathbf{T}}, \dots, \mathbf{T}^{(p-1)}; \mathbf{F}, \dot{\mathbf{F}}, \dots \mathbf{F}^r), \tag{10.14}$$

where $\mathbf{g}$ is a tensor-valued function, and (10.14) is subject to the initial data

$$\mathbf{T}(t_0), \dot{\mathbf{T}}(t_0), \dots \mathbf{T}^{(p-1)}(t_0), \tag{10.15}$$

so that (10.14) has a unique solution. A *material of integral type* is one where the response functional F can be expressed in terms of an integral polynomial of an appropriate tensor representation of the deformation gradient $\mathbf{F}$. For example, one can use the function

$$\mathbf{G}^*(s) = \mathbf{C}_{(t)}^*(t - s) - \mathbf{1}, \tag{10.16}$$

to describe the deformation history, where $\mathbf{G}^*(s) \equiv \mathbf{0}$ corresponds to the rest history and $\mathbf{C}*_{(t)}$ is related to the right Cauchy–Green tensor $\mathbf{C} = \mathbf{U}^2 = \mathbf{F}^T\mathbf{F}$. For more complicated flow conditions, where electric, magnetic or chemical effects need to be taken into account, and for more complex materials with microstructure, the constitutive relation for $\mathbf{T}$, or $\mathbf{q}$, should also depend on appropriate variables, such as temperature, volume fraction, fabric, electric and magnetic fields, etc.

From a continuum perspective, there are many different approaches which are used to develop the stress tensor. There are those studies which focus on the onset of the flow, or the yield condition, and therefore techniques of theory of plasticity have been used. Theories of structured continua have also been used; for example, Kanatani [49] uses the theory of micropolar fluids for flowing granular materials. In this theory, for example, the velocity and the rotational velocity of particles are used as two independent kinematic field variables. The similarity between rapid flow of granular materials and turbulent flows was observed by Blinowski [11] and as a result many modified forms of turbulent flow theories have been developed (for example, Massoudi and Ahmadi [61] developed a constitutive theory for rapid flows of granular materials, where the effects of density and fluctuation energy gradients are taken into account). The rapid flow is generally maintained by particle collisions causing the particles to have highly irregular paths; these irregular motions create fluctuations on all field variables such as velocity, temperature, etc. Ogawa [78] and Ogawa et al. [79], recognizing the discrete nature of granular materials, defined two different kinds of temperature: one is the usual temperature associated with the thermal fluctuation of the molecules of each grain, and the other is related to the "random" translational and rotational fluctuations of the individual grain. This indicates a similarity to the kinetic theory of gases, and as a result after the publication of these papers many different kinds of kinetic theory of granular materials have been developed. The key reason behind using a kinetic theory approach for granular materials is the analogy between a grain and a molecule. As a first approximation, both gas molecules and granular materials can be considered single-sized hard spheres. If the flow of granular materials is rapid, it can be characterized by an average or a continuum velocity about which the velocities of the individual grains change because of random collisions. Since such a motion closely resembles a flowing gas, the ideas incorporated into the kinetic theory of gases have been applied to the rapid flow of granular materials. However, the analogy between a gas and granular materials is not exact because the collisions between gas particles are energy conserving, whereas those between grains are energy dissipating. But the inelasticity of granular collisions prevents the theoretical methods developed for hard sphere gases from being directly applied to flowing granular materials. To calculate the macroscopic behaviour of a flowing gas, only the probability of any two particles simultaneously occupying two given positions and exhibiting two specified velocities is needed; furthermore, if the fluid is dilute enough, all that is needed is the probability of any one particle being at a given position with a specified velocity. These ideas are more fully explained by utilizing the concept of a phase space [67]. Boyle and Massoudi [12] derived a constitutive equation by utilizing the ideas of Enskog's dense gas theory; their model can predict normal-stress differences arising due to density gradient.[1] It is important to point out that in theories for rapid flow

---

[1] We would like to mention that the application of kinetic theory of gases to flows of granular materials is plagued by many assumptions, perhaps beyond what the original theory may have stood for. In the last two decades, the researchers in the field of granular materials have exploited the techniques of kinetic theory and statistical mechanics. There are certain flow regimes where the collisions of particles are rare, in the sense that the flow is so slow and the particles are so densely packed that one cannot assume the basic assumptions in the kinetic theory are valid [116]. However, certain processes such as fluidized beds present a special challenge: before the onset of

of granular materials based on a kinetic theory approach, the fluctuations in the velocity field give rise to the notion of "granular temperature". The convective heat transport, within the context of such theories, is determined by the fluctuations in the velocity field, usually known as the pseudo-energy equation. Application of kinetic theory of gases to solid granular particles suffers from the fact that a theory based on such assumptions can only at best, if proper considerations and measures are taken, describe the flow behaviour of granular materials when the particle velocity is tremendously high, there are collisions of particles taking place, and the concept of mean free path also exists. In the continuum approach that we have adopted here, the fluctuations in the velocity field are ignored, and the phenomenon of heat transfer is incorporated in the energy equation. To include in addition to the energy equation, the notion of granular temperature would be inconsistent with our approach. The continuum approach is applicable when the packing of the material is reasonably compact, i.e., high volume fraction, and the fluctuations from the mean are not significant. In the kinetic theory approach, additional boundary conditions are also necessary for the value of the fluctuating energy, which is related to what is usually referred to as the "granular temperature". The convective heat transport, within the context of such theories, is determined by the fluctuations in the velocity field, usually known as the pseudo-energy equation. Application of kinetic theory of gases to solid granular particles suffers from the fact that a theory based on such assumptions can only at best, if proper considerations and measures are taken, describe the flow behaviour of granular materials when the particle velocity is tremendously high, there are collisions of particles taking place, and the concept of mean free path also exists. In the continuum approach that we have adopted here, the fluctuations in the velocity field are ignored, and the phenomenon of heat transfer is incorporated in the energy equation. To include in addition to the energy equation, the notion of granular temperature would be inconsistent with our approach. The continuum approach is applicable when the packing of the material is reasonably compact, i.e., high volume fraction, and the fluctuations from the mean are not significant. In the kinetic theory approach, additional boundary conditions are also necessary for the value of the fluctuating energy, which is related to what is usually referred to as the granular temperature.

The effect of boundaries on the flow of granular materials has been studied [21, 32, 34, 35, 38, 104] Whether one uses the continuum approach or the kinetic theory approach, slip may often occur at the wall, especially when the interstitial fluid is a gas, and therefore the classical assumption of adherence boundary condition at the wall no longer applies. Lugt and Schot [54] give a review of slip

---

fluidization, the flow regime is perhaps more in the slow deformation range and after fluidization in the rapid flow range. It is in the rapid flow regime that the kinetic theory approach may be used. Thus, we tend to look at the kinetic-theory based models as tools, which may be appropriate for some cases and irrelevant or not appropriate in other cases. Just as one can build a bridge or design a ship without ever having access to the tools in statistical theories, it is also possible to study many engineering problems involving granular materials without ever using the kinetic theory approach.

flow. While the phenomenon of slip at the wall occurs more frequently in the flow of rarefied gases and certain polymers, for the majority of fluid flows, the no-slip boundary condition is a reasonable one.

Many researchers have advocated that the stress tensor is composed of two parts: a rate-independent part, similar to a yield condition, for the frictional flow regime, and a rate-dependent part for the viscous flow regime [46, 47, 82, 83] . At some point, these various theories may overlap and it is not always possible to model the materials accordingly. Most granular materials exhibit two unusual and peculiar characteristics: (1) normal stress differences, and (2) yield criterion. The first was observed by Reynolds [97, 98] who called it "dilatancy." Dilatancy is described as the phenomenon of expansion of the voidage that occurs in a tightly packed granular arrangement when it is subjected to a deformation. Many of the existing theories for the flowing granular materials use this observation to relate the applied stress to the voidage and the velocity. Reiner [94,95] proposed and derived a constitutive relation for wet sand whereby the concept of dilatancy is given a mathematical structure. This model does not take into account how the voidage (volume fraction) affects the stress. However, using this model, Reiner showed that application of a non-zero shear stress produces a change in volume. The constitutive relation of the type

$$S_{nm} = f_0 \delta_{nm} + 2\eta D_{nm} + 4\eta_c D_{nj} D_{jm}, \tag{10.17}$$

describing the rheological behaviour of a non-linear fluid was named by Truesdell [117] as the Reiner–Rivlin [99] fluid, where in modern notation the stress tensor $\mathbf{T}$ is related to $\mathbf{D}$ [1, 8, 14]:

$$\mathbf{T} = -p(\rho)\mathbf{I} + f_1 \mathbf{D} + f_2 \mathbf{D}^2, \tag{10.18}$$

where f_1 and f_2 are functions of ρ, $tr\mathbf{D}$, $tr\mathbf{D}^2$. This is an early example of one of the constitutive formulations where the concept of normal stress differences is discussed.

The second peculiarity is that for a granular material to flow, there is often a yield stress below which the particles do not undergo deformation. The yield condition is often related to the angle of repose, friction and cohesion among other things. Perhaps, the most popular yield criterion for granular materials is the Mohr-Coulomb one, although by no means the only one [63]. Overall, it seems that the phase that the bulk solids most resemble, at least macroscopically, is that of a non-linear (non-Newtonian) fluid, specifically a visco-plastic one. The science of studying non-linear fluids is "Rheology" and according to Reiner [96]: "Rheology started when Bingham [9] investigated concentrated clay-suspensions, and Bingham and Green [10] investigated oilpaints." In fact, Bingham [9] proposed a constitutive relation for a visco-plastic material in a simple shear flow where the relationship between the shear stress (or stress $\mathbf{T}$ in general), and the rate of shear (or the symmetric part of the velocity gradient $\mathbf{D}$ is given by [81]

$$2\mu D_{ij} = \begin{Bmatrix} 0 & \text{for} & F < 0 \\ F T'_{ij} & \text{for} & F \geq 0 \end{Bmatrix}, \tag{10.19}$$

where T_{ij}^t denotes the stress deviator and F, called the yield function, is given by

$$F = 1 - \frac{K}{{\Pi_2'}^{\frac{1}{2}}},$$ (10.20)

where Π_2' is the second invariant of the stress deviator, and in simple shear flows it is equal to the square of the shearing stress and K is called yield stress [2] (a constant).

Bagnold [7] performed experiments on neutrally buoyant, spherical particles suspended in Newtonian fluids undergoing shear in co-axial rotating cylinders. He distinguished three different regimes of flow behaviour, which he termed macro-viscous, transitional and grain-inertia. In the so-called "macro-viscous" region, which corresponds to low shear rates, the shear and normal stresses are linear functions of the velocity gradient. In this region, the fluid viscosity is the dominant parameter. In the "grain-inertia region", the fluid in the interstices does not play an important role and the dominant effects arise from particle–particle interactions. Here, the shear and the normal stresses are proportional to the square of the velocity gradient. Connecting the two limiting flow regimes was the Bagnold's transitional flow, in which the dependence of the stress on shear rate varied from a linear one corresponding to the macro-viscous regime to a square dependence predicted for the grain-inertia flow regime. The interesting phenomenon was the presence of a normal stress proportional to the shear stress, similar to that of the quasi-static behaviour of a cohesionless material obeying the Mohr–Coulomb criterion [105, 107]. Many investigators have modelled flow of particles as a fluid [5, 29, 33, 68, 102, 112] . For a review, see the article by Elaskar and Godoy [24] .

A model which, in theory, is capable of yield stress and normal stress effects is that of Rajagopal and Massoudi [86] where the Cauchy stress tensor **T** depends on the manner in which the granular material is distributed, i.e., the volume fraction v, its gradient and the symmetric part of the velocity gradient tensor **D** (See also [19, 20]):

$$\mathbf{T} = [\beta_0(v) + \beta_1 \nabla v \cdot \nabla v + \beta_2 \text{tr } \mathbf{D}]\mathbf{1} + \beta_3 \mathbf{D} + \beta_4 \nabla v \otimes \nabla v + \beta_5 \mathbf{D}^2,$$ (10.21)

where

$$\mathbf{D} = \frac{1}{2}\left[\nabla \mathbf{u} + (\nabla \mathbf{u})^T\right],$$ (10.22)

where $\rho = \gamma v$, and γ is constant. The material parameters $\beta_0 - \beta_5$ are assumed to have the following forms:

$$\beta_0 = kv, \quad k < 0,$$ (10.23)

---

[2] For one-dimensional flow, these relationships reduce to the ones proposed by Bingham [9], i.e.,
$F = 1 - \frac{K}{|T_{12}|}$, and $2\mu D_{12} = \begin{cases} 0 & \text{for } F < 0 \\ F T_{12}' & \text{for } F \geq 0 \end{cases}$. The constitutive relation is known as Bingham model (See also [123], p.170, or Zhu et al. [122]).

$$\left\{\begin{aligned}
\beta_1 &= \beta_1{}^*(1 + v + v^2) \\
\beta_2 &= \beta_2{}^*(v + v^2) \\
\beta_3 &= \beta_3{}^*(v + v^2) \\
\beta_4 &= \beta_4{}^*(1 + v + v^2) \\
\beta_5 &= \beta_5{}^*(1 + v + v^2)
\end{aligned}\right\}. \tag{10.24}$$

The above representation can be viewed as Taylor series approximation for the material parameters [88]. Such a quadratic dependence, at least for the viscosity β_3, is on the basis of dynamic simulations of particle interactions [119, 120]. Further restrictions on the coefficients have been obtained by using the following argument [56]. Since the stress should vanish as $v \to 0$, we can conclude that

$$\beta_{50} = \beta_{30} = \beta_{20} = 0. \tag{10.25}$$

This is a principle of the limiting case. That is, if there are no particles, then v and $grad\,v$ are zero, and the stress should be zero; however, the kinematical terms $\mathbf{D}$, $\mathbf{D}^2$ and $tr\,\mathbf{D}$, multiplied by β_2, β_3 and β_5 do not necessarily go to zero when there are no particles. Therefore, to ensure this we impose the above restrictions. Furthermore, Rajagopal and Massoudi and Rajagopal et al. [86, 87] have shown that

$$k < 0, \tag{10.26}$$

as compression should lead to densification of the material. They gave the following rheological interpretation to the material parameters: $\beta_0(v)$ is similar to pressure in a compressible fluid and is to be given by an equation of state, $\beta_2(v)$ is like the second coefficient of viscosity in a compressible fluid, $\beta_1(v)$ and $\beta_4(v)$ are the material parameters connected with the distribution of the granular materials, $\beta_3(v)$ is the viscosity of the granular materials, and $\beta_5(v)$ is the coefficient of cross viscosity. Looking at (10.21) with (10.23) and (10.24), it can be shown that this model is capable of predicting both normal stress differences and yield stress [63]. In our analysis, for simplicity, we assume β_5 is zero and as a result only one of the normal stress differences is non-zero. For a discussion on experimental measurements of these material properties, see Rajagopal et al. [90] and Baek et al. [6].

The constitutive relation for the heat flux vector is assumed to be given by the Fourier's law of conduction where the heat flux vector $\mathbf{q}$ is linearly related to the temperature gradient:

$$\mathbf{q} = -K\nabla\theta, \tag{10.27}$$

where K is an effective or modified form of the thermal conductivity. The dependence of thermal conductivity on material properties and volume fraction is one of the challenging problems in mechanics of microstructure materials. There have been some studies with regard to porous materials but few with regard to flowing granular materials. For example, Kaviany [50] presents a thorough review of the appropriate correlations for the thermal conductivity for packed beds and the effective thermal conductivity in multiphase flows. Massoudi [59, 60] has recently given a brief review of this subject and has also proposed and derived a general constitutive relation for the heat flux vector for a flowing granular media. A general and non-linear model presented by Jeffrey [45], which includes the second-order effects in the volume fraction is given by

$$K = K_m[1 + 3\beta v + \hat{\beta}v^2] + O(v^3), \tag{10.28}$$

where

$$\hat{\beta} = 3\beta^2 + \frac{3\beta^3}{4} + \frac{9\beta^3}{16}\left(\frac{\alpha+2}{2\alpha+3}\right) + \frac{3\beta^4}{26} + \cdots, \tag{10.29}$$

where

$$\beta = \frac{\alpha-1}{\alpha-2}, \tag{10.30a}$$

$$\alpha = \frac{k_2}{k_1}, \tag{10.30b}$$

where α is the ratio of conductivity of the particle to that of the matrix, K the effective conductivity of the suspension, K_m the conductivity of the matrix and v is the solid volume fraction.

10.5 Boundary Value Problem: *Free Surface Flow Down an Inclined Plane*

Most of the constitutive relations used in mechanics, whether non-Newtonian models, turbulence models, etc., when substituted in the general governing equations, i.e., the balance laws, would produce a system of partial differential equations which at times are impossible to solve completely with the numerical techniques currently available. Therefore, from a modelling point of view, it is worthwhile to study problems where due to simplification of the kinematics of the flow or the boundary conditions, one obtains a system of (non-linear) ordinary differential equations. The solution to these simpler problems would be useful for at least two different reasons: (1) they provide insight into the nature of these non-linear constitutive relations, and (2) they provide examples where the accuracy or convergence of solutions to the general multi-dimensional equations can be tested. Other interesting phenomena such as stability and uniqueness of solutions also sometimes arise. Furthermore, the higher order terms in the constitutive relations require additional boundary conditions, and this itself is an important problem to study.

Flow of granular materials down an inclined plane occurs naturally as in the cases of avalanches and mudslides; it is also used for transporting and drying of bulk solids (such as agricultural and pharmaceutical products). It is a viscometric flow [115] and one which amends itself to fundamental theoretical and experimental studies.

Fig. 10.3 The idealized geometry of flow over an inclined surface

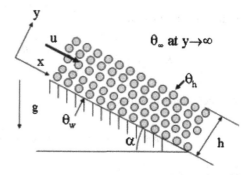

Hutter et al. [40, 41] showed that the existence (or non-existence) of solutions for fully developed flow down an inclined plane depends on the type of boundary conditions that are imposed. Rajagopal et al. [87] also studied existence and uniqueness of solutions to the equations for the flow of granular materials down an inclined plane, where the thermal effects were also included. They used the model developed by Rajagopal and Massoudi [86]; in that study, they delineated a range of values for the material parameters, which ensures existence of solutions to the equations under consideration. They also proved rigorously that for a certain range of values of the parameters no solutions exist, while for others there is a multiplicity of solutions. In many problems, viscous dissipation is ignored. There are many cases in polymer rheology and lubrication, for example, where viscous dissipation cannot be neglected [27, 110, 111]. For densely packed granular materials, as particles move and slide over each other, heat is generated due to friction and therefore in such problems, the viscous dissipation should be included. It is assumed that the flow is fully developed (see Fig. 10.3). The free surface is exposed to high ambient temperature and as a result a modified Stefan–Boltzmann correlation [27] for radiation is used at that surface (For a detailed discussion of this problem, see [64]).

We make the following assumptions:

1. The motion is steady
2. The effects of radiant heating r are imposed at the free surface
3. The constitutive equation for the stress tensor is given by (10.21)–(10.26) and the constitutive equation for the heat flux vector is that of Fourier's assumption, given by (10.27)–(10.30)
4. The density (or volume fraction), velocity and temperature fields are of the form

$$\left\{ \begin{array}{l} v = v(y) \\ \mathbf{u} = u(y)\mathbf{i} \\ \theta = \theta(y) \end{array} \right\}. \tag{10.31}$$

With the above assumptions, the equation of conservation of mass is automatically satisfied and the dimensionless forms of the momentum and the energy equations become:

y-direction momentum:

$$R_1 \frac{dv}{d\overline{y}} + R_2(1 + v + v^2)\frac{dv}{d\overline{y}}\frac{d^2v}{d\overline{y}^2} + \frac{R_2}{2}(1 + 2v)\left(\frac{dv}{d\overline{y}}\right)^3 = v\cos\alpha, \quad (10.32)$$

x-direction momentum:

$$R_3 v(1 + v^2)\frac{d^2\overline{u}}{d\overline{y}^2} + R_3(1 + 2v)\frac{dv}{d\overline{y}}\frac{d\overline{u}}{d\overline{y}} = -v\sin\alpha. \quad (10.33)$$

Heat Transfer:

$$\left(1 + av + bv^2\right)\frac{d^2\overline{\theta}}{d\overline{y}^2} + (a + 2bv)\frac{dv}{d\overline{y}}\frac{d\overline{\theta}}{d\overline{y}} = -R_4 v(1 + v)\left(\frac{d\overline{u}}{d\overline{y}}\right)^2, \quad (10.34)$$

where the dimensionless quantities $\overline{y}$ (distance), the velocity $\overline{u}$ and the temperature θ are given by the following equations [73]:

$$\overline{y} = \frac{y}{h}, \quad \overline{u} = \frac{u}{u_0}, \quad \overline{\theta} = \frac{y}{\theta_w}, \quad (10.35)$$

where u_o is a reference velocity, θ_w is the wall temperature and h is the constant height to the free surface, and where

$$R_1 = \frac{k}{\gamma g h}, \quad R_2 = 2\frac{(\beta_1{}^* + \beta_4{}^*)}{\gamma g h^3}, \quad R_3 = \frac{\beta_3{}^* u_w{}^2}{2h^2\gamma g},$$

$$R_4 = \frac{\beta_3{}^* u_w{}^2}{2K_m\theta_w}, \quad R_5 = \frac{\varepsilon\sigma\theta_w{}^3 h}{K_m}. \quad (10.36)$$

The dimensionless parameters have the following physical interpretations: R_1 could be thought of as the ratio of the pressure force to the gravity force; R_2 is the ratio of forces developed in the material, due to the distribution of the voids, to the force of gravity; R_3 is related to the Reynolds number; R_4 is a measure of viscous dissipation, which is the product of the Prandtl number and the Eckert number; and R_5 is a measure of the emissivity of the particles to the thermal conductivity. It follows from Rajagopal and Massoudi [86] that R_1 must always be less than zero for the solution to exist and R_3 and R_4 must be greater than zero, since the viscosity is positive. In addition to these dimensionless numbers, values for N, a, b and α are to be given a priori (10.38a, b).

From (10.32), it is clear that we need two boundary conditions for the volume fraction v and (10.33) and (10.34) indicate that two conditions are required for the velocity and the temperature fields, respectively. We can also see that (10.32) is not coupled to the other two equations, and it can therefore be integrated first. Once the volume fraction field is determined, (10.33) can be solved for "u", and finally (10.34) is integrated to find the temperature field θ.

At the surface of the incline, we assume the no-slip condition for the velocity and a constant temperate θ_w:

$$at \quad \overline{y} = 0 : \begin{cases} \overline{u} = 0 \\ \theta = 1 \\ v = v_0 \end{cases}.$$ (10.37)

At the free surface, the no-traction boundary condition is imposed on the stress tensor, and as a result we obtain two expressions for the velocity gradient and the volume fraction (10.38a, b), and for the temperature we apply the Stefan–Boltzmann[3] condition [100] when the surrounding temperature is designated as θ_∞ and the temperature at the free surface is θ_h. Thus, we have

$$at \quad \overline{y} = 1 : \begin{cases} \frac{d\overline{u}}{dy} = 0, & \text{(a)} \\ R_1 v + \frac{R_2}{2}\left(\frac{dv}{dy}\right)^2 = 0, & \text{(b)} \\ \frac{d\overline{\theta}}{dy} = \frac{-R_5}{1+av+bv^2}\left(\overline{\theta}_{\overline{y}=1}^4 - \overline{\theta}_\infty{}^4\right); & \text{where } \overline{\theta}_\infty = \frac{\theta_\infty}{\theta_w}. \quad \text{(c)} \end{cases}$$

(10.38)

For v there are at least two options: (1) we can impose a distribution function for v at the wall, which could have a constant value (this may mean glueing particles to the surface), or (2) we can give an average value for v integrated over the cross section (a measure of the amount of particles present in the system),

$$\overline{N} = \int_0^1 v\,dy.$$ (10.39)

We will use condition (10.39) for the volume fraction.

The objective is to conduct a parametric study to see how the various parameters $(R_1, R_2, R_3, R_4$ and $R_5)$ affect the solution. The volume fraction equation has a non-unique solution as discussed by Rajagopal et al. [31, 87]. Such a non-unique solution stems from the boundary condition at $y = h$ by (10.38b), where one can see that for a given value of the volume fraction at $y = h$ there exist two possible conditions of the derivative of the volume fraction: one is negative and the other is positive. Thus, there must be two solutions to satisfy these conditions: one solution in which v must increase monotonically from $\overline{y} = 0$ to $\overline{y} = 1$ to so that the positive condition, $\left(\frac{dv}{dy}\right)_{\overline{y}=1} < 0$ is satisfied and one in which must decrease monotonically to satisfy the negative condition at the free surface, $\left(\frac{dv}{dy}\right)_{\overline{y}=1} > 0$. Typical results of such multiple solutions are shown in Fig. 10.4. However, due to gravitational

---

[3] Equation (10.38c) is really our first approximation and a more appropriate one for the case of granular materials might be to introduce into the equation a function for the dependence of the volume fraction, for example,

$$q = f(v)\varepsilon\sigma(\theta_h{}^4 - \theta_s{}^4).$$

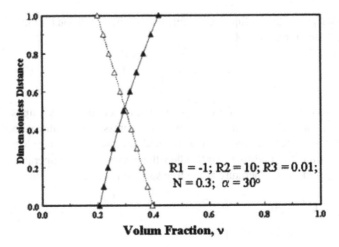

Fig. 10.4 Multiple solutions of the volume fraction equation

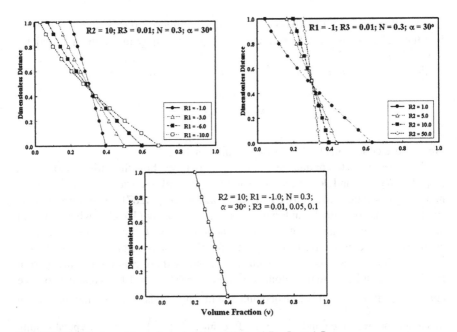

Fig. 10.5 Distribution of the volume fraction: Effects of R_1, R_2 and R_3

effect, particles must settle down towards the surface of the wall rather than float upwards to the free surface. For this reason, the solution in which v decreases as one approaches the free surface is chosen as the correct solution.

The effects of parameters R_1, R_2 and R_3 on v are shown in Fig. 10.5. The distribution of the volume fraction, however, is independent of parameter R_3. For example, by keeping other parameters constant the volume fraction decreases from

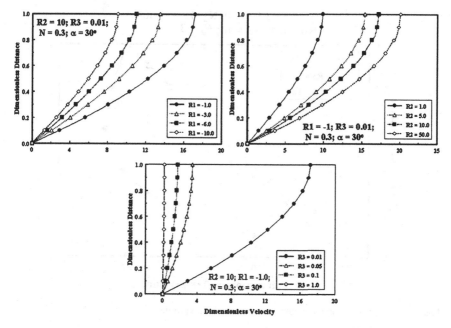

Fig. 10.6 Dimensionless velocity as a function of the dimensionless distance: effects of R_1, R_2 and R_3

0.355 at the inclined plane to 0.2 at the free surface for all values of R_3 from 0.01 to 1.0. Since R_3 represents the viscous forces, R_1 the pressure force and R_2 the material forces due to normal stress effects, this result indicates that the distribution of the volume fraction is dominated by the material forces and the pressure rather than the viscous forces. The effects of R_1 , R_2 and R_3 on the velocity profiles are shown in Fig. 10.6. The velocity profile has a parabolic shape with the maximum value at the free surface. The value of the free surface velocity increases as R_2 increases but it decreases when R_1 increases. Although R_3 does not have a significant effect on v, it has a strong influence on the velocity distribution. The results indicate that as R_3 increases the flow slows down and the velocity profile approaches a straight line.

For the temperature calculations, we chose $a = 0.75$ and $b = 1$. The effect of R_3 on the temperature distribution is shown in Fig. 10.7 with the surrounding temperature being higher than the surface temperature, $\theta_\infty = 3.5$. For $R_3 = 0.01$, the maximum temperature is found to be located in a region within the flow and the temperature decreases as one approaches the inclined plane and the free surface. This is the effect of viscous heat generation. As R_3 increases, the temperature profile becomes linear, increasing monotonically from the wall towards the free surface. This is because the flow slows down as R_3 increases as shown in Fig. 10.6, and the viscous dissipation term and the convection term in the energy equation are not as significant as the conduction term. In this case, the particles are heated primarily through conduction.

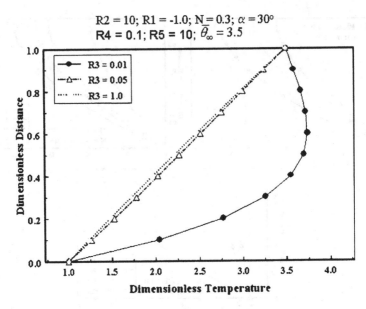

Fig. 10.7 Effect of R_3 on the dimensionless temperature profile

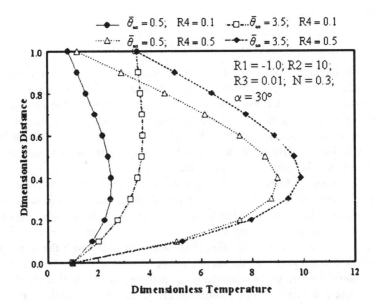

Fig. 10.8 Effect of R_4 on the dimensionless temperature profile

The effects of R_4 on the temperature are shown in Fig. 10.8, where the temperature profiles and the free surface temperature are calculated as a function of R_4 while other parameters are kept constant. Two different cases are investigated: the cooling case, i.e., when the temperature of the environment is lower than the surface temperature and the heating case, i.e., when the environment temperature is higher than the surface temperature. An increase in R_4 means that the heat generated by viscous dissipation term is increased. As a result, this generated heat must be transferred to the surface and be removed from the free surface through radiation. Increasing R_4 therefore increases both the maximum temperature in the flowfield and the free surface temperature. For example, when the environment is hotter than the free surface, as R_4 increases, the free surface temperature increases slightly. When the environment temperature is lower than the surface temperature, an increase in R_4 results in a significant increase in the free surface temperature. For all cases, the maximum temperature and the free surface temperature are always higher than both the surface and the surrounding temperatures.

The effect of R_5 on the temperature profile for cooling and heating cases is shown in Fig. 10.9. For both cases, the maximum temperature is found to be inside the flowfield and heat is transferred toward both the inclined surface and the free stream. This is primarily due to the viscous dissipation effect which raises the flowfield temperature to a higher value than both the surface temperature and the environment temperature. The parameter R_5 represents the heat transfer (due to radiation) at the free surface and its magnitude is a measure of how much of the heat generated by the viscous dissipation term can be transferred away. Thus, an increase in R_5 leads to a decrease in the free surface temperature until it reaches the environment temperature. When this condition is reached, the solution becomes independent of R_5. The free surface temperature could be less than or greater than the inclined surface temperature and the environment temperature depending on the competition between R_4 and R_5. For a constant R_4, if R_5 is small the heat generated due to

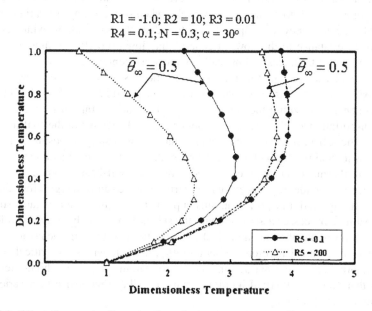

Fig. 10.9 Effect of parameter R_5 on the dimensionless temperature profile

viscous dissipation cannot be sufficiently removed by the radiation process. In this case, the free surface temperature is higher than both the inclined surface and the environment temperature. If R_5 is large, however, the radiative heat removal can sufficiently balance the heat generated. In this case, the free surface temperature is low and it decreases as R_5 increases. If R_5 continues to increase, the free surface temperature will reach the surrounding temperature.

10.6 Concluding Remarks

In processes such as (dense phase) flow of particles in a bin, hopper, silo, etc., where only solid constituents are involved, i.e., it is assumed that the effects of the interstitial fluid are negligible, it is essential to develop models to accurately describe the material characteristics, especially the rheological properties. This need is further strengthened by the fact that a better and economically efficient process would require a design which is not only based on an accurate constitutive representation for the materials involved but also on the various other factors which govern the flow characteristics of the granular materials. In addition, moisture content of the bulk solids is an important factor controlling the flow properties of the granular materials; the surface moisture leads to the appearance of cohesive forces between particles of solids and of adhesive forces between particles and the walls of the container. Both retard the flow of solids and under certain conditions may stop the flow entirely. Since for the same weight of solid the total surface of solids is greater for small grains, the surface moisture content increases inversely as the particle diameter. For that reason, fine particles display more cohesive and adhesive forces than the larger grain solids. Furthermore, fine particles when stored for a certain time undisturbed have a tendency to compact that is to reduce the total volume which creates additional resistance to the flow. In general, the flow properties of most materials can be expected to decrease drastically as moisture content increases, particularly for finer materials [108]. When a coal stockpile is stored in the presence of air, slow oxidation of the carbonaceous materials occurs and heat is released. If the rate of heat generation within the stockpile is greater than the rate of heat dissipation and transportation to the external environment, the self-heating of the coal stockpile ensues. The self-heating of coal stockpiles has a long history of posing significant problems to coal producers because it lowers the quality of coal and may result in hazardous thermal runaway. Precise prediction of the self-heating process is, therefore, necessary to identify and evaluate control measures and strategies for safe coal mining, storage and transportation. Such a prediction requires an accurate estimate of the various processes associated with the self-heating, which are impossible unless the appropriate phenomenological coefficients are known. In such storage-type problems, the critical ignition temperature θ_{cr}, also known as the critical storage temperature, is an important design and control parameter, since at higher temperatures than this θ_{cr}, thermal ignition occurs, possibly giving rise to a variety of instabilities and problems.

References

1. Adams MJ, Briscoe BJ, Kamjab M (1993) The deformation and flow of highly concentrated dispersions. Adv Colloid Interface Sci 44:143–182
2. Ahmadi GA (1982) Generalized continuum theory for granular materials. Int J Non-Linear Mech 17:21–33
3. Antman, SS (1995) Nonlinear problems of elasticity. Springer-Verlag, New York
4. Antony SJ, Hoyle W, Ding Y (2004) Granular materials: Fundamentals and applications. The R Soc of Chemistry, Cambridge, UK
5. Astarita T, Ocone R (1999) Unsteady compressible granular materials. Ind Eng Chem Res 38:1177–1182
6. Baek S, Rajagopal KR, Srinivasa AR (2001) Measurements related to the flow of granular materials in a torsional rheometer. Part Sci Technol 19:175–186
7. Bagnold RA (1954) Experiments on a gravity free dispersion of large solid spheres in a Newtonian fluid under shear. Proc R Soc Lond 225:49
8. Batra RC (2006) Elements of continuum mechanics. American Institute of Aeronautics and Astronautics (AIAA) Inc., Reston, VA
9. Bingham EC (1922) Fluidity and plasticity. McGraw-Hill, New York
10. Bingham EC, Green H (1919) Testing materials II. American Assoc 19:610
11. Blinowski A (1978) On the dynamic flow of granular media. Arch Mech 30:27–34
12. Boyle EJ, Massoudi M (1990) A theory for granular materials exhibiting normal stress effects based on Enskog's dense gas theory. Int J Eng Sci 28:1261–1275
13. Brown RL, Richards JC (1970) Principles of powder mechanics. Pergamon Press, London
14. Caswell B (2006) Non-Newtonian flow at lowest order, the role of the Reiner-Rivlin stress. J Non-Newt Fluid Mech 133:1–13
15. Chapman S, Cowling TG (1990) The mathematical theory of non-uniform gases. Cambridge University Press, Cambridge, MA
16. Collins IF, Houlsby GT (1997) Applications of thermomechanical principles to the modeling of geotechnical materials. Proc R Soc Lond A 453:1975–2001
17. Collins IF (2005) Elastic/plastic models for soils and sands. Int J Mech Sci 47:493–508
18. Coussot P (2005) Rheometry of pastes, suspensions, and granular materials. Wiley, Hoboken, NJ
19. Cowin SC (1974) A theory for the flow of granular material. Powder Tech 9:61–69
20. Cowin SC (1974) Constitutive relations that imply a generalized Mohr-Coulomb criterion. Acta Mech 20:41–46
21. Craig K, Buckholtz RH, Domoto G (1987) The effects of shear surface boundaries on stresses for the rapid shear dry powders. ASME J Tribol 109:232
22. de Gennes PG (1998) Reflections on the mechanics of granular matter. Physica A 261: 267–293
23. Duran J (2000) Sands, powders and grains. Springer, New York
24. Elaskar SA, Godoy LA (1998) Constitutive relations for compressible granular materials using non-Newtonian fluid mechanics. Int J Mech Sci 40:1001–1018
25. Fan LS, Zhu C (1998) Principles of gas–solid flows. Cambridge University Press, Cambridge
26. Fang C, Wang Y, Hutter K (2006) A thermo-mechanical continuum theory with internal length for cohesionless granular materials. Part I. A class of constitutive models. Continuum Mech Thermodyn 17:545–576
27. Fuchs HU (1996) The dynamics of heat. Springer-Verlag Inc., New York
28. Gidaspow D (1994) Multiphase flow and fluidization. Academic Press, San Diego
29. Goddard JD (1984) Dissipative materials as models of thixotropy and plasticity. J Non-Newt Fluid Mech 14:141
30. Goldhirsch I (2003) Rapid granular flows. Annu Rev Fluid Mech 35:267–293
31. Gudhe R, Rajagopal KR, Massoudi M (1994) Heat transfer and flow of granular materials down an inclined plane. Acta Mech 103:63–78

32. Gutt GM, Haff PK (1991) Boundary conditions on continuum theories of granular material flow. Int J Multiphase Flow 17:621
33. Haff PK (1983) Grain flow as a fluid-mechanical phenomenon. J Fluid Mech 134:401–430
34. Haff PK (1986) A physical picture of kinetic granular fluids. J Rheol 30:931
35. Hanes DM, Inman DL (1985) Observations of rapidly flowing granular–fluid materials. J Fluid Mech 150:357
36. Hermann HJ (1999) Statistical models for granular materials. Physica A, 263:51–62
37. Hermann HJ, Luding S (1998) Modeling granular media on the computer. Continuum Mech Thermodyn 10:189–231
38. Hui K, Haff PK, Unger JE et al (1984) Boundary conditions for high-shear grain flows. J Fluid Mech 145:223–233
39. Hutter K (1983) Theoretical glaciology. Kluwer, Hingham, MA
40. Hutter K, Szidarovszky F, Yakowitz S (1986) Plane steady shear flow of a cohesionless granular material down an inclined plane: A model for flow avalanches Part I. Acta Mech 63:87–112
41. Hutter K, Szidarovszky F, Yakowitz S (1986) Plane steady shear flow of a cohesionless granular material down an inclined plane: A model for flow avalanches Part II: Numerical Results. Acta Mech 63:239–261
42. Hutter K, Rajagopal KR (1994) On the flows of granular materials. Continuum Mech. Thermodyn 6:81–139
43. Jaeger HM, Nagel SR, Behringer RP (1996) Granular solids, liquids, and gases. Rev Modern Phys 68:1259
44. Jackson R (2000) The dynamics of fluidized particles. Cambridge University Press, Cambridge, MA
45. Jeffrey DJ (1973) Conduction through a random suspension of spheres. Proc R Soc Lond A 335:355–367
46. Johnson PC, Jackson R (1987) Frictional–collisional constitutive relations for granular materials with application to plane shearing. J Fluid Mech 176:67–93
47. Jyotsna R, Kesava Rao K (1997) A frictional-kinetic model for the flow of granular materials through a wedge-shaped hopper. J Fluid Mech 346:239–270
48. Kalman H, Tardos GI (2005) Elements of particle technology in the chemical industry. Particulate Sci Tech 23:1–19
49. Kanatani KI (1979) A micropolar continuum theory for the flow of granular materials. Int J Eng Sci 17:419–432
50. Kaviany M (1995) Principles of heat transfer in porous media. 2nd edn. Springer, New York
51. Klausner Y (1991) Fundamentals of continuum mechanics of soils. Springer, New York
52. Liu IS (2002) Continuum mechanics. Springer, Berlin
53. Lu SY, Kim S (1990) Effective thermal conductivity of composites containing spheroidal inclusions. AIChE J 36:927–938
54. Lugt HJ, Schot JW (1974) A review of slip flow in continuum physics. In: Lugt HJ (ed) Proc Second Symp Fluid–Solid Surface Interactions. Naval Research and Development Center, Bethesda, MD
55. Marcus RD, Leung LS, Klinzing GE et al (1990) Pneumatic conveying of solids. Chapman and Hall, London
56. Massoudi M (2001) On the flow of granular materials with variable material properties. Int J Non-Linear Mech 36:25
57. Massoudi M (2004) Constitutive modelling of flowing granular materials: A continuum approach. In: Antony SJ, Hoyle W, Ding Y (eds) Granular materials: Fundamentals and applications. The Royal Society of Chemistry, Cambridge, UK, 63–107
58. Massoudi M (2005) An anisotropic constitutive relation for the stress tensor of a rod-like (fibrous-type) granular material. Math Probl Eng 679–702
59. Massoudi M (2006) On the heat flux vector for flowing granular materials, Part I: Effective thermal conductivity and background. Math Methods Appl Sci 29:1585–1598
60. Massoudi M (2006) On the heat flux vector for flowing granular materials, Part II: Derivation and special cases. Math Methods Appl Sci 29: 1599–1613

61. Massoudi M, Ahmadi GA (1994) Rapid flow of granular materials with density and fluctuation energy gradients. Int J Non-Linear Mech 29:487–492
62. Massoudi M, Boyle EJ (2001) A continuum-kinetic theory approach to the flow of granular materials: The effects of volume fraction gradient. Int J Non-Linear Mech 36:637–648
63. Massoudi M, Mehrabadi MM (2001) A continuum model for granular materials: Considering dilatancy, and the Mohr-Coulomb criterion. Acta Mech 152:121–138
64. Massoudi M, Phuoc TX (2006) Boundary value problems in heat transfer studies of granular materials modeled as compressible non-linear fluids. Math Probl in Eng Article ID 56046 pp. 1–31. doi:10.1155/MPE/2006/56046
65. Massoudi M, Phuoc TX (2007) Conduction and dissipation in the shearing flow of granular materials modeled as non-Newtonian fluids. Powder Technol 175:146–162
66. Maugin GA (1999) The thermomechanics of nonlinear irreversible behaviors. World Scientific Publishing Co, River Edge, NJ
67. McQuarrie DA (1976) Statistical mechanics. Harper & Row Publishers, New York
68. McTigue DF (1982) A non-linear constitutive model for granular materials: Applications to gravity flow. J Appl Mech 49:291–296
69. Mehrabadi MM, Nemat-Nasser S, Oda M (1982) On statistical description of stress and fabric in granular materials. Int J Numer Anal Methods Geomech 6:95–108
70. Mehrabadi MM, Loret B, Nemat-Nasser S (1993) Incremental constitutive relations for granular materials based on micromechanics. Proc R Soc Lond A 441:433–463
71. Mehta A (ed) (1994) Granular matter. Springer, New York
72. Müller I (1967) On the entropy inequality. Arch Rat Mech and Anal 26:118–141
73. Na TY (1979) Computational methods in engineering boundary value problems. Academic Press, New York
74. Nedderman RM (1992) Statics and kinematics of granular materials. Cambridge University Press, Cambridge
75. Nemat-Nasser S, Mehrabadi MM (1984) Micromechanically based rate constitutive description for granular materials. In: Desai CS, Gallagher RH (eds) Mechanics of engineering materials. Wiley, 451–463
76. Nozad I, Carbonell RG, Whitaker S (1985) Heat conduction in multiphase systems-I: Theory and experiment for two-phase systems. Chem Eng Sci 40:847–855
77. Oda M (1972) The mechanism of fabric change during compressional deformation of sand. Soils Foundations 12:1–18
78. Ogawa S (1978) Multitemperature theory of granular materials. In: Cowin SC, Satake M (eds) Proc U.S.–Japan Seminar on Continuum Mechanical and Statistical Approaches in the Mechanics of Granular Materials. Sendai, Japan, 208–217
79. Ogawa S, Umemura A, Oshima N (1980) On the equations of fully fluidized granular materials. J Appl Math Phys (ZAMP) 31:483–493
80. Plasynski SI, Peters WC, Passman SL (1992) The department of energy solids transport, multiphase flow program. Proc NSF–DOE Joint Workshop on Flow of Particulates and Fluids, Gaithersburg, September 16–18
81. Prager W (1989) Introduction to mechanics of continua. Dover Publications, Inc, Mineola, NY
82. Prakash JR, Kesava Rao K (1988) Steady compressible flow of granular materials through a wedge-shaped hopper: The smooth wall, radial gravity problem. Chem Eng Sci 43:479–494
83. Prakash JR, Kesava Rao K (1991) Steady compressible flow of cohesionless granular materials through a wedge-shaped bunker. J Fluid Mech 225:21–80
84. Rajagopal KR (2003) On implicit constitutive theories. Appl Math 48:579–319
85. Rajagopal KR (2006) On implicit constitutive theories for fluids. J Fluid Mech 550:243–249
86. Rajagopal KR, Massoudi M (1990) A method for measuring material moduli of granular materials: Flow in an orthogonal rheometer. Topical Report, DOE/PETC/TR-90/3
87. Rajagopal KR, Troy WC, Massoudi M (1992) Existence of solutions to the equations governing the flow of granular materials. Eur J Mech B/Fluids 11:265–276
88. Rajagopal KR, Massoudi M, Wineman AS (1994) Flow of granular materials between rotating disks. Mech Res Comm 21:629–634

89. Rajagopal KR, Tao L (1995) Mechanics of mixtures. World Scientific Publishing. River Edge, NJ
90. Rajagopal KR, Gupta G, Yalamanchili RC (2000) A rheometer for measuring the properties of granular materials. Part Sci Tech 18:39–55
91. Rajagopal KR, Srinivasa AR (2004) On thermomechanical restrictions of continua. Proc R Soc Lond A 460:631–651
92. Ranade VV (2002) Computational flow modeling for chemical reactor engineering. Academic Press, San Diego
93. Rao K, Kesava, Nott PR (2008) An introduction to granular flow. Cambridge University Press, New York
94. Reiner M (1945) A mathematical theory of dilatancy. Am J Math 67:350–362
95. Reiner M (1948) Elasticity beyond the elastic limit. Am J Math 70:433–466
96. Reiner M (1958) Rheology. In: Flugge S (ed) Handbuch Der Physik, Vol. VI. Springer, Berlin
97. Reynolds O (1885) On the dilatancy of media composed of rigid particles in contact with experimental illustrations. Phil Mag Series 5(20):469–481
98. Reynolds O (1886) Experiments showing dilatancy, a property of granular material, possibly connected with gravitation. Proc R Inst GB 11:354–363
99. Rivlin RS (1948) The hydrodynamics of non-Newtonian fluids. I Proc R Soc Lond 193: 260–281
100. Saldanha da Gama RM (2004) On the conduction/radiation heat transfer problem in a body with wavelength-dependent properties. Appl Math Model 28:795–816
101. Schotte W (1960) Thermal conductivity of packed beds. AIChE J 6:63–67
102. Savage SB (1979) Gravity flow of cohesionless granular materials in chutes and channels. J Fluid Mech 92:53–96
103. Savage SB (1984) The mechanics of rapid granular flows. Adv Appl Mech 24:289–366
104. Savage SB, Sayed M (1984) Stress developed by dry cohesionless granular materials in an annular shear cell. J Fluid Mech 142:391–430
105. Schaeffer DG (1987) Instability in the evolution equations describing incompressible granular flow. J Diff Eq 66:19–50
106. Soo SL (1990)Multiphase fluid dynamics. Science Press, Brookfield
107. Spencer AJM (1982)Deformation of ideal granular materials. In: Mechanics of solids. Pergamon Press, Oxford and New York 607–652
108. Stepanoff AJ (1969) Gravity flow of bulk solids and transportation of solids in suspension. Wiley, New York
109. Straughan B, Greve R, Ehrentraut H et al (eds) (2001) Continuum mechanics and applications in geophysics and the environment. Springer, Berlin
110. Szeri AZ (1998) Fluid film lubrication. Cambridge University Press, Cambridge, MA
111. Szeri AZ, Rajagopal KR (1985) Flow of a non-Newtonian fluid between heated parallel plates. Int J Non-Linear Mech 20:91
112. Tardos GI (1997) A fluid mechanistic approach to slow, frictional flow of powders. Powder Technol 92:61–74
113. Tardos GI, McNamara S, Talu I (2003) Slow and intermediate flow of a frictional bulk powder in the Couette geometry. Powder Technol 131:3–39
114. Torquato S (1987) Thermal conductivity of disordered heterogeneous media from the microstructure. Rev Chem Eng 4:151
115. Truesdell C (1976) The meaning of viscometry in fluid mechanics. Annu Rev Fluid Mech 6:111–146
116. Truesdell C, Muncaster RG (1980) Fundamentals of Maxwell's kinetic theory of a simple monatomic gas. Academic Press, New York
117. Truesdell C, Noll W (1992) The non-linear field theories of mechanics. Springer, New York
118. Tsotsas E, Martin H (1987) Thermal conductivity of packed beds: A review. Chem Eng Process 22:19–37
119. Walton OR, Braun RL (1986) Stress calculations for assemblies of inelastic spheres in uniform shear. Acta Mech 63(1–4):73–86

120. Walton OR, Braun RL (1986) Viscosity, granular-temperature, and stress calculations for shearing assemblies of inelastic, frictional disks. J Rheol 30:949–980
121. Zhang X, Jeffrey RG, Mai YW (2006) A micromechanical-based Cosserat-type model for dense particulate solids. Z Angew Math Phys (ZAMP) 57:682–707
122. Zhu H, Kim YD, De Kee D (2005) Non-Newtonian fluids with a yield stress. J Non-Newt Fluid Mech 129:177–181
123. Ziegler H (1983) An introduction to thermomechanics. 2nd revised edn. North-Holland, Amsterdam

Author Index

Subject Index